The Beachcomber's Gu

Michael Stachowitsch

The Beachcomber's Guide to Marine Debris

Michael Stachowitsch
Department of Limnology and Bio-Oceanography
University of Vienna
Vienna, Austria

ISBN 978-3-319-90727-7   ISBN 978-3-319-90728-4 (eBook)
https://doi.org/10.1007/978-3-319-90728-4

Library of Congress Control Number: 2018941090

© Springer International Publishing AG, part of Springer Nature 2019
This work is subject to copyright. All rights are reserved by the Publisher, whether the whole or part of the material is concerned, specifically the rights of translation, reprinting, reuse of illustrations, recitation, broadcasting, reproduction on microfilms or in any other physical way, and transmission or information storage and retrieval, electronic adaptation, computer software, or by similar or dissimilar methodology now known or hereafter developed.
The use of general descriptive names, registered names, trademarks, service marks, etc. in this publication does not imply, even in the absence of a specific statement, that such names are exempt from the relevant protective laws and regulations and therefore free for general use.
The publisher, the authors, and the editors are safe to assume that the advice and information in this book are believed to be true and accurate at the date of publication. Neither the publisher nor the authors or the editors give a warranty, express or implied, with respect to the material contained herein or for any errors or omissions that may have been made. The publisher remains neutral with regard to jurisdictional claims in published maps and institutional affiliations.

Printed on acid-free paper

This Springer imprint is published by the registered company Springer Nature Switzerland AG
The registered company address is: Gewerbestrasse 11, 6330 Cham, Switzerland

Foreword

There is no shortage of terrible news in the world. It is easy to be overwhelmed with a century's weight of human folly. Hailed as humanity's liberator, the industrial age has exacted a high toll. Our deep addiction to fossil fuels and all the products that stem from petroleum and coal has accelerated climate change, poisoned waterways, and polluted every ocean and sea. Industries have flourished by misjudging the ocean as an inexhaustible space, one that could swallow and absorb our refuse within its depths. The Earth's wondrous blue ecosystems have become our dumping ground for metal, glass, plastic, wood, paper, fabrics, and oil.

When I founded OceanCare it was with a deep passion for all of Earth's oceanic creatures, from the mightiest whales to the tiniest plankton. The extraordinary fluid network covering seventy percent of the Earth's surface – oceans, seas, estuaries, lagoons, mangroves, coral reefs, salt marshes, intertidal zones, and the sea floor – inspired our work. We knew that marine life determined the very nature of our planet. Ocean ecosystems are our lungs, our sustenance, and the force that keeps our climate stable. Without living oceans, we would cease to exist.

Beaches are where most of us connect with oceans. They are the backdrop of idyllic holidays, or the scene of dreams we hold for escape. Water lapping on the shore is deeply calming, and holding a shell freshly plucked from the sand evokes the mysteries of another life lived somewhere unfamiliar. Even when filled with people, or lashed by winds, the beach holds a soothing power as the water coils around our feet and sand moves between our toes. Standing at the water's edge, looking out to sea, Earth feels unique, boundless, and beautiful.

Despite our fantasies, the ocean depths are not empty and their volume is finite. In my years heading conservation campaigns I have watched grim news for oceans roll towards the shore in waves, amassing like seaweed on a beach after a storm.

Decades ago, when plastic litter was first recognized as a looming problem, the world community hoped it would remain contained to plastic islands far out to sea where it might be collected. Its greatest threat appeared to be for marine species ingesting plastic bags. Occasionally metal, glass, or wood washed ashore, but our imaginations still saw these as treasures returned from the deep, worn smooth and polished by grains of sand and salt. Even our phrases romanticised what we saw – *driftwood, flotsam,* and *jetsam.*

As each year progressed, the magnitude of the problem amplified. Today, almost every shore has been degraded by litter, every lagoon tarnished by debris. Almost no reef is unsullied with floating plastic. The scale can be overwhelming. This constant onslaught of dreadful news disrupts our thinking and changes our brain, keeping us in a state of stress and threatening to paralyze action. This mindset does nothing to change the crisis.

Humans are ingenious beings, and I have faith we can find a way out of this problem. I know that compassionate information, underpinned by solutions, empowers us and strengthens our faith in humanity. When facts are packaged with wry humour, the message can motivate us to make changes in our own lives. This is the gift of *The Beachcomber's Guide to Marine Debris*.

Delivering this crucial information in the form of a nature guide is inspired. Historically, such guides were designed to tap into our desire for escape – they were a map to the wilds. We can still go there. We can still either walk the beaches in our minds or with our feet genuinely heated by the sand, but now we travel with purpose. *The Beachcomber's Guide to Marine Debris* enlightens us towards powerful change.

Sigrid Lüber
Founder and President – OceanCare
www.oceancare.org

Preface

Beaches are among the most magical environments on our planet, and strolling barefoot on the sand is certainly one of life's great pleasures. This may help explain why so much of the world's population lives directly on the coast; in the United States alone, half the population lives within 50 miles of the seashore. And the beach is a preferred vacation destination for nearly everyone.

One of the strongest attractions of beaches is the prospect of experiencing pristine, untamed nature. Today, however, our shorelines are changing beyond the natural give and take between land and waves. Pollution and habitat degradation are probably the first things that come into mind when we think about the sea. One of the most visible and pervasive types of marine pollution is the trash that clutters our waterways and beaches. You may call it trash, rubbish, litter, garbage, refuse, junk, or flotsam and jetsam, but specialists refer to it as marine debris. This is not merely an aesthetic problem. Marine scientists, along with a wide range of private organizations, government agencies, and international bodies, now recognize marine debris as a scourge for sustainable tourism, a serious threat to wildlife, a menace to fisheries and boating, a health hazard to humans, and an economic threat to coastal communities.

Today's beachcombers are more likely to encounter man-made debris than shore animals and plants. This book takes on the challenge: it combines the world's most beloved tourist destinations (beaches) with their most visible threat. Designed along the lines of a traditional nature guide, it picks up where the others left off. Rather than showcasing algae, seashells, and other shore dwellers, it covers the items that actually dominate our beaches, lake shores, and river banks today: plastic, glass, metal, wood, paper, oil, apparel, and other materials in unending variations, combinations, and unsavory states of decomposition. Whereas traditional nature guides are tailored to regional faunas and floras, you can take this book anywhere in the world. This is because pollution – and marine debris in particular – is a global problem. The world's oceans are all interconnected. International corporations, standardized products and packaging, worldwide sales and distribution – coupled with winds, currents, and waves – leave recognizable litter strewn over the seven seas and their coastlines.

Whether you are participating in a beach cleanup, wish to identify a curious item you (almost) stepped on, or merely need a more thought-provoking and entertaining beach vacation read, this book is the answer. It will tell you what you found, where it came from, how and when it might decompose, point to potential hazards, and suggest alternative products or waste reduction and prevention (recycling and upcycling) ideas. What we are doing to our seas and shores is a crime, and beaches around the world are crime scenes – the visible manifestation of our impact. Fortunately, marine debris or beach litter is one type of pollution that we as individuals can actually do something about, both as consumers and as beachcombers. So get into the action, become a beach detective, have a little fun in the process, and be part of the solution in this barefoot environment.

Vienna, Austria   Michael Stachowitsch

Acknowledgment

I thank my editor Janet Slobodien and Springer for taking on this book, which clearly transcends traditional boundaries and challenged a clear fit into many an academic publishing program. Two very positive anonymous reviewers helped make it happen. I am especially grateful to the OceanCare team, above all Niki Entrup, for their enthusiasm about this book and for supporting its production. Petra Triessnig, coauthor in the first paper on beach litter and sea turtle hatchlings, was instrumental in organizing my untold photos. The geographic distribution of the photos reflects the length of my sojourns on various shores more than any country's proclivity for producing marine debris. I am thankful to the people who didn't arrest me for taking compromising photos of their polluted beaches – even if most had no inkling of what I was actually up to. Martin Zuschin provided Fig. 3.47 and Fabian Ritter Figs. 3.48 and 3.71. Alexandra Haselmair drew the four graphics. I wish to thank my wife, Sylvie, and the many friends and colleagues who badgered me to complete this book, including Coni Dennig who helped hatch the idea on a vacation to Italy all those years ago. This book was written in the spirit of love for seascapes near and far.

Contents

1	**Introduction**	1
	1.1 Guide Format	1
	1.2 The Beach Environment	3
	1.3 The Marine Debris Problem	6
	1.3.1 Marine Debris: Why the Fuss?	7
	1.3.2 Gaining a Better Understanding of Marine Debris	11
	1.4 What Can You and I Do?	22
	References	28
2	**Glass**	31
	2.1 Glass	31
	2.2 Glass Bottles and Fragments	33
	2.3 Light Bulbs	44
	References	48
3	**Metal, Vehicles, and Tires**	49
	3.1 Metal	49
	3.2 Vehicles	68
	3.3 Tires	76
	References	86
4	**Plastic**	87
	References	95
	4.1 Plastic Beverage Containers and Co.	96
	References	99

4.2 Plastic Canisters	114
4.3 Toys	121
4.4 Balloons and Co.	129
References	129
4.5 Household Plastic	134
4.6 Plastic Bags and Other Packaging	144
4.7 Shotgun Shells	154
5 Foamed Plastic (Styrofoam)	**159**
References	162
6 Hygiene	**171**
6.1 Personal Hygiene	171
6.2 Toilets and Co.	183
References	189
7 Medical Wastes	**191**
References	193
8 Furniture and Furnishings	**201**
8.1 Beach Furniture	201
8.2 Appliances and Home Electronics	212
References	217
9 Apparel	**219**
9.1 Clothing	219
9.2 Footwear	226
9.2.1 Shoes	226
9.2.2 Flip-Flops	234
9.3 Gloves	241
9.4 Hats and Caps	244
References	248
10 Water Sports	**249**

11	**Fishing Gear**	263
	References	277
12	**Wood**	279
	12.1 Boats and Household	279
	12.2 Pallets	291
	References	299
13	**Paper**	301
	References	312
14	**Organic Wastes**	313
	References	327
15	**Oil and Tar**	329
	References	336
16	**Smoking**	337
	References	351
Index		353

About the Author

Michael Stachowitsch received his B.Sc. degree at the University of Pittsburgh and his Ph.D. at the University of Vienna. He has successfully managed to maintain a career as a marine biologist in a landlocked country, conducting research in the Adriatic and Red Seas and teaching numerous university courses. He is the long-term Austrian coordinator of a Mediterranean sea turtle conservation project in Turkey and represents Austria at the International Whaling Commission. Michael's many activities as a scientist, author, translator, and editor have taken him to beaches around the globe: the fauna and flora differ from shore to shore, but the beach litter is astoundingly uniform – and abundant. Decades on hands and knees with a camera yielded this book and revealed unsettling truths about human attitudes and behavior.

What better place to contemplate the state of the world's beaches than at the scene of the crime (homage to Frank Zappa). A recent international beach cleanup yielded 56 toilets.

1

Introduction

1.1 Guide Format

Virtually every type of item ever produced by humans has been found discarded on some beach somewhere. You can find everything *including* the kitchen sink (Fig. 3.49). Categorizing and presenting this wide range of products are a difficult task in any context – managing it within the framework of a practical field guide is daunting. Nonetheless, the globalization of markets and universality of many products; the general economic principles behind producing, selling, buying, and discarding goods; and the commonalities of human behavior help make this book valid and applicable worldwide.

The book's format parallels that used in many nature guides. Marine debris or beach litter is divided into a system of hierarchic categories. The main chapters are Glass, Metal, Plastic, Foamed Plastic, Hygiene, Medical, Furniture and Furnishings, Apparel, Water Sports, Fishing Gear, Wood, Paper, Organic Wastes, Oil and Tar, and Smoking. These 15 chapters are then broken down into more specific subcategories. Glass, for example, is split into Bottles, Light Bulbs, and Glass Pieces. Apparel contains Clothing, Footwear, Gloves, and Hats. Plastic is subdivided into Beverage Containers, Canisters, Toys, Balloons, Bags and Packaging, and Shotgun Shells.

Each main chapter and subcategory is introduced with a short text providing information on why these items are found on beaches and how they may have gotten there, their synonyms and composition, life expectancy in the beach environment, relative abundance, and how to interpret various labels, symbols, and pictograms. The introductions also include the potential hazards

Fig. 1.1 Each bagful of beach picnic garbage tends to produce most major marine debris categories: paper, glass, plastic, metal, and organic wastes. Mediterranean, Turkey

posed to wildlife or humans, tips for proper handling during cleanups, and potential reduction or recycling ideas.

Categorizing marine debris is further complicated by the fact that many items are composed of more than one material. Light bulbs, at least initially, consist of the glass body itself and a metal fitting. A shotgun shell consists mostly of a plastic shell and plastic wad but also of a metal base. Moreover, trash deposited on beaches after visits tends to contain a full mix of components (Fig. 1.1). Wherever possible, the items are categorized based on their main component. If more logical or practical, however, the chapters are based on the use category rather than material composition. Accordingly, syringes – which can consist of a glass or plastic tube, a metal needle, and perhaps a rubber plug – are placed in the chapter Medical. Apparel can range from swimsuits to hats, shoes and flip-flops, or gloves, each made of different materials. This warrants a separate chapter. See the Table of Contents for a quick overview of the different chapters and the products they contain, or find your item quickly in the Index.

This guide is designed to:

- Draw attention to the incredible spectrum of items polluting our beaches, lake shores, and riverbanks.
- Help you identify such items. In many cases, they are instantly recognizable, in others not. This is why various decomposition stages are highlighted.

- Enable you to quickly and accurately fill out survey data forms if you are participating in a beach cleanup.
- Allow you, as an advanced beachcomber or "beach detective," to better interpret individual debris items on site.
- Provide a deeper insight into marine and beach pollution in general.
- Point out potential hazards to wildlife and to yourself.
- Suggest waste-reducing ideas, more ecologically-friendly alternative products, or recycling/upcycling possibilities.

The photographs are the backbone of this guide. Most items were photographed as found directly on the beach by the author without prior manipulation.

The selection of items is based on those that are either most likely to be encountered, especially interesting or unexpected, or that represent particular hazards. The instant brand recognition perfected by marketing experts means ever-more-extravagant and unique product and packaging designs: they can never be too large, too glossy, or too fancy and gaudy. This is eminently useful to help "garbologists" or, in the present case, beachcombers to recognize most items quickly. Key decomposition stages or breakdown components are presented to help you identify even worn or unusual debris articles or fragments. Often a trace of color pattern, a hint of original shape, or vestiges of other design features or lettering are sufficient to identify an item. Based on what might be called the "law of marine debris accumulation" – the likelihood that more of the same or related items can be found lying nearby – the chances are good that you'll succeed with an identification based on better-preserved specimens. Soon you'll be able to recognize items from afar or even if only a small part is projecting from the sand. A walk down your local supermarket aisles or a quick Internet search can help refresh your memory about the original products and provide names and addresses in case you or your team want to take further action by contacting manufacturers, local retail stores, authorities, or your legislators both local and national.

1.2 The Beach Environment

The beach is a truly magical environment, but I probably don't need to tell you that because you have either been there, are preparing to head that way, or have already spread your towel out on the sand. People go to the beach for many different reasons. Some are surfers or fishers, and others come for a picnic or simply to lie on the warm sand and dig into a summer read. Others

Fig. 1.2 Are we so out of tune about beaches as natural environments that we have to be told how to behave? Atlantic, USA

come because their kids love to dig holes and build sand castles – and can be loud and rambunctious without creating havoc at home or school. Still others visit to experience nature, to get away from it all and regenerate. And some folks actually work there, from life guards and beach bar owners to sea turtle biologists. All of us, however, are at least partially trapped in a trend reflecting modern life: it might be referred to as an increasing disconnect between ourselves and nature and wildlife (Fig. 1.2).

Let's take a closer look at the "nature" part. Beaches, as the interface between land and sea, are unique, living environments – living both in the sense that they harbor distinct animal and plant communities and because they are constantly changing, ever on the move thanks to wind, waves, and currents. While this may sound perfectly logical, natural, and acceptable to you, it often irks the powers to be and other influential stakeholders who want "their" beach to remain in the same state and configuration as they first experienced it or as they may otherwise "need" it. We humans crave a certain environmental stability that promotes the orderly development of our affairs. Unfortunately, the seashore can be very wild and rugged indeed, and virtually every effort we have ever made to tame it and achieve a semblance of stability has gone awry. Every breakwater, pier, or other structure that we place into the surf or on the beach has typically had an unexpected and unintended effect or has shifted the "problem" a short stretch up or down the coast, to the next-door neighbors, the next beach concession, and the adjoining city limits or county line.

Let's go back to the other "nature" aspect of living beaches, namely the different habitats and their inhabitants. Experts differentiate numerous zones along beaches, from the shallow-water zones up to the border to fully terrestrial vegetation [1]. This can include, from bottom to top, more common names such as the tidal zone, surf zone, swash zone, foreshore, backshore, and dune landscape. Specialists sometimes use a variety of more technical terms such as sublittoral, infralittoral, mid-littoral, supralittoral, etc. The width of the zones can vary widely depending on the slope of the shore or the extent of the tides. Each has its own physical and environmental conditions, namely wetness/dryness, exposure to wave action, particle size, and salinity. Each also has its own fauna and flora. On sandy shores, but even more so on rocky ones, the corresponding zonations may also be named after the dominant plant and animal groups living there. On rocky shores, brown algae, mussels, or barnacles sometimes form clearly visible bands.

Protecting the shoreline and the fauna and flora of its different zones has become a priority issue in many places. In the USA, the quarterly journal *Shore & Beach* is devoted entirely to this mission [2]. Often, however, localized efforts have gone awry because we fail to understand or fully factor in the interconnectedness of these many beach zones and adjoining habitats. Allowing coral reefs to become degraded, for example, exposes the coastline to storm waves and accelerated erosion. Cutting down mangrove forests alters the sediment flows and nutrient conditions of all neighboring ecosystems, exposing reefs to terrestrial inputs. Destroying shallow-water seagrass meadows by dredging, anchoring, or destructive fishing methods alters water turbulence and sand stability, not to mention animal communities. Landward, sensitive dune landscapes with their fragile vegetation are destroyed not only by all-terrain vehicles but even by our own footprints, a phenomenon known as trampling, requiring major restoration efforts [3]. On beaches themselves, efforts to reverse damage or restore eroding beaches often wreak even more havoc: erecting wind fences or replenishing the beach with sand sucked up a short distance offshore destroys seafloor communities and piles sand up, often in the wrong composition, amounts, and position. As we all know, it's not nice to fool Mother Nature, and many such attempts quickly backfire.

Whatever your motivation for going to the beach, one of the most commonly heard disappointments is that the beach is dirty. And by dirt most folks mean beach litter or marine debris – items that we ourselves have produced and whose final resting place on beaches is often directly or indirectly our doing. What parts of the beach are affected by marine debris? The answer is you'll probably be able to find it in all the abovementioned zones. Typically, the freshest accumulations – along with seaweeds and other organic material

Fig. 1.3 Beach cleanups must include the backshore areas, where many people deposit their garbage when leaving and where much of the windblown beach litter becomes trapped or snagged in the vegetation. Mediterranean, Turkey

known as beach wrack – are deposited at the high tide mark or along one of the smaller so-called drift lines that mark former high tides or the successive progress of the receding water. Wind and storm waves, however, ultimately distribute this material far and wide. Heavy wood and metal items often come to rest further up the beach at the foot of elevated beach features such as foredunes or in depressions. Glass and plastic bottles often roll away until they hit the first obstacles or steeper slopes. The lightest materials such as paper and plastic bags are quickly blown up the beach, where they can be found snagged and draped in the dune vegetation or in fences (Fig. 1.3). During cleanups, this means spreading the teams out to cover these many zones, both to include the full expanse of the beach and to find all the variously distributed debris.

1.3 The Marine Debris Problem

Before we get into any details about marine debris, its types and categories, and why we need to make a fuss about it, we need a workable definition. It's fair to begin by saying that many terms are floating about – marine debris, marine litter, beach litter, marine anthropogenic litter, or ocean trash. This depends a bit on your perspective, but they all essentially refer to the same phenomenon. The United Nations Environment Programme (UNEP) defines it as "any persistent, manufactured or processed solid material discarded, disposed of or abandoned in the marine and coastal environment" [4]. One practical translation of that might be "ugly stuff that we put there but wouldn't want to step on."

What we are talking about is pollution, clear and simple. Traditionally, the impact of human civilization ended at the waterfront. This is no longer the case today. Our relationship with the oceans has always been volatile, and humans have historically been on the receiving end, bearing the brunt of nature's caprices. In the meantime, however, we have increasingly turned the tide and imposed our – mostly disruptive – whims on nature. Accordingly, when talk turns to "humankind and the sea," our first association these days is often marine pollution. The oceans, as the world's largest ecosystem, receive a large share of our wastes. In fact, most types of pollution turn out to be marine pollution: in the long run, the wastes we produce reach the sea by way of rivers and the atmosphere. Direct ocean dumping adds insult to injury. Marine debris has become a hot scientific research topic. It is the topic of international symposia [5] and special volumes [6], and a good percentage of all articles in the premier marine pollution journal *Marine Pollution Bulletin* are currently devoted to it [7]. The recognition that beachcombers are increasingly likely to encounter marine debris is reflected in separate chapters in new guides on seashore animals and plants [8], and a first booklet devoted entirely to plastic marine debris has been written for the German North Sea and Baltic Sea (in German) [9]. This is only the tip of the proverbial iceberg because we see merely the visible material floating on the water surface or washed ashore. Especially high concentrations have accumulated in the swirling gyres in the middle of the major oceans (catchword: Great Pacific Garbage Patch), much like sugar accumulates in the middle of your stirred teacup. This has even prompted a public relations gag in which the UN is being petitioned to recognize the "Trash Isles" as a new country, complete with passports and celebrity honorary citizens [10]. Much also settles on the seafloor, where it is visible only to divers, scientists operating underwater robots or seated in submersibles, and fishers who dredge it up with their nets. This guide emphasizes the debris most accessible to the beachcomber – items cast ashore or left there by visitors. Importantly, rubbish in general and in the present marine debris context tells us quite a bit about ourselves [11], explaining why archeologists have long used buried waste heaps to reconstruct societies throughout human history.

1.3.1 Marine Debris: Why the Fuss?

Why worry about beach debris when most people have come to accept litter as a fact of life? Garbage in the environment, whether on the street or on the shore, is the symptom of a larger problem. It represents the visible spectrum of more subtle pollutants and, as such, is a barometer for total pollution levels. The vast amounts of trash produced by modern civilization have become a key environmental issue. The crisis is evident on all levels – chaos during disruptions in garbage collection, overflowing landfills, overtaxed incinerators, and large-scale shipping of garbage to other countries. This is reflected in mounting obstacles when selecting toxic waste sites, in grass roots "not-in-my-backyard" movements, and in increasing outrage over international trafficking in industrial wastes.

The result is an environment smothered in trash. No ever so remote part of the globe has been spared. Special expeditions are being organized to the Himalayan mountains to remove the debris accumulated by decades of mountaineering. Polar research stations are being ostracized for their garbage disposal practices. Even outer space has become a repository of trash: about 30,000 items from nuts and bolts to spent satellites and rockets (and more recently an automobile from a public relations stunt) clutter Earth's orbit and pose a serious threat to future space flight.

Why do beaches merit special attention? For one, most marine pollution takes place in shallow coastal waters, whether on purpose or accidentally. This is where most fisheries are concentrated, where oil and gas exploration and exploitation take place. Tourism and recreational boating are concentrated in these waters, sewage pipes end there. Direct dumping into the ocean – everything from construction material to nuclear wastes – has also proven to be more convenient closer to shore. It is only a matter of time before much of this material reaches the beach. Beaches merely mirror the burden of the overall biosphere. The attention given to beach pollution can also be explained on a more personal level. Litter on city streets and in parks raises little more than an eyebrow these days: encountering trash in environments that we expect to be pristine usually elicits a howl of protest. The oceans, and by extension the seashore, traditionally represent such a vast, untamed environment. Here, where we open ourselves up to experience nature, any confrontation with the evils of civilization is all the more disappointing and disturbing.

Today's beachcomber is more likely to be confronted with trash than with the natural elements of the ecosystem. This is equally true for remote islands as it is for heavily populated tourist beaches. In fact, remote, unpopulated islands – whose shorelines are not regularly cleaned – often show the world's highest accumulations of marine debris and serve as reference beaches for this form of pollution [12]. The problem is not merely esthetic. Among the many types of marine pollution, marine debris is increasingly being recognized as a major threat to marine wildlife and to humans themselves. To name just a few, it entangles, dismembers, strangulates, and drowns marine mammals and turtles, blocks the digestive tracts of seabirds and fish, smothers organisms on the seafloor, fills fishermen's nets, clogs the cooling systems of boats, and poses a health hazard to bathers, beachcombers, and seafood consumers. Wow! Perhaps less known is the threat posed by the organisms encrusting or otherwise hanging on to floating debris. Although on an evolutionary scale this phenomenon helps explain the colonization and development of life on remote islands, the problem has gotten out of hand today. The hitch-hiking species include bryozoans, barnacles, polychaete worms, hydroids, crabs, and mollusks, to name a few [13]. These organisms use such "floating hotels" to raft thousands of miles to faraway habitats. This brings exotic, "alien" species to places they normally would never reach, where they can outcompete and displace native species, significantly disturbing ecosystem balance and posing a major threat to global biodiversity [14].

And if the esthetic and biological arguments fail to hold sway, then we need to recruit the dirtiest of all issues, the one argument that grabs everyone's attention – money. Marine debris at sea and on beaches is a costly matter. If your boat engine's cooling intake system becomes clogged with plastic bags and overheats, you'll need to completely overhaul or replace the most expensive item on the vessel. Hitting large floating marine debris (let's say a lost ship container) can mean the loss of your boat (and your life). More generally, the investment of coastal municipalities in beach-cleaning equipment and personnel can be enormous, but cleaning operations are a major intrusion into the "living" beach ecosystem (Fig. 1.4). Finally, heavily polluted beaches are a

Fig. 1.4 Heavy-duty machinery to help clean up marine debris replaces one evil with another. Beyond missing the smaller litter, this "manicuring" destroys any remaining semblance of beaches as natural ecosystems. Atlantic, Morocco

reason for not going there. This translates into lost income for seaside communities. After a large amount of marine debris washed up on the beaches of one tourist island after a heavy rainfall, the visitor count dropped by 63%, resulting in a tourism revenue loss of between US$29 and 37 million [15]. That's the type of information that can prod even the least conservation-minded authorities into taking action.

In fact, the issue has become so urgent that a full range of private organizations, citizens' groups, businesses, and government agencies on the local and regional scale, along with nongovernmental organizations (NGOs) and international governmental organizations (IGOs) up to an including the United Nations have risen to the challenge and are taking action.

You too can join the good fight. This is because, as opposed to contamination by heavy metals, pesticides, and other chemicals or radioactivity, you as an individual can actually do something about this form of pollution. The first step is conscientious behavior on the beach to eliminate one significant source of debris: littering. You can also help clean up beaches: most types of marine debris can be removed without prior training and technical aids (see subchapter "What Can You and I Do?"). This book will take you through the first steps, namely, recognizing and correctly classifying refuse on beaches, pointing out sources and threats; alternative, more environmentally friendly products; and approaches to reducing the amount of discardable material in the first place. We need to recognize that "business as usual" will leave our beaches buried under growing mountains of trash.

The UNEP definition given above puts emphasis on so-called persistent marine debris. This means material that is durable and decomposes only slowly. For most people this is synonymous with plastic, which forms much of the debris on beaches. The same physical features that have allowed plastics to replace traditional materials are also responsible for their longevity and negative impact on the environment. Nonetheless, plastics are by no means the only persistent debris items with long-term environmental effects. Glass and metal, for example, may take decades and centuries to decompose. Many essentially degradable materials are also more long-lived than most folks assume [16]. This includes paper and wood, which are usually specially treated in some way (laminated paper or cardboard, impregnated or painted wood). Food items such as fruits with big seeds or whose outer surfaces are tough, waxed, or sprayed (oranges, banana peels, etc.) may take a long time to decompose. Of course, short-lived doesn't mean harmless. A wide range of short-lived unsavory items (e.g., those flushed down toilets) may be particularly unesthetic, hazardous, symptomatic of failed waste treatment, or indicative for the presence of other less visible types of pollution. All are regarded as marine debris and treated as such in this guide.

1.3.2 Gaining a Better Understanding of Marine Debris

There are many ways of getting a better grasp on marine debris. The first step is to recognize the range of items and the different ways of categorizing them. The second step is to examine the many sources and think about solutions.

One way to classify marine debris is based on size. Accordingly, researchers distinguish a range beginning with nano- and microplastics (the latter smaller than 1 or 5 mm, depending on your perspective) all the way up to so-called mega-litter (larger than 2–3 cm according to some researchers but going up to many meters in length: think derelict fishing gear). The individual pieces are variously referred to as objects, items, articles, pieces, or fragments, but we don't need to go into this any further because scientists themselves have not agreed on a standardized terminology.

Other criteria for categorizing marine debris include the source of the pollution (e.g., land-based versus ocean-based), the branch of industry that produces the original product, the function of the item, or the composition of the material. The present guide uses a combination of composition and function, with emphasis on material composition. The 15 main categories are glass, metal, plastic, foamed plastic, hygiene, medical, furniture, apparel, water sports, fishing gear, wood, paper, organic wastes, oil and tar, and smoking. This breakdown is an amalgamation of the classification used in international beach cleanups and in various guidelines for conducting marine debris surveys [17].

Examining the sources is the step toward finding solutions and reducing the inputs. Waste enters the sea from many points of origin and by various routes. A common distinction is between land- and ocean-based sources.

Land-Based Sources

Land-based sources are also sometimes referred to as onshore sources and include shoreline and recreational activities. This means beachgoers themselves, direct pollution by coastal industries, inputs from further inland via streams and rivers, street runoff, sewage outflows and storm sewer overflows, solid waste disposal facilities, and illegal or inappropriate dumps.

River-borne wastes are a prime indirect source of marine debris. Did you ever wonder why so many industries are heavily concentrated around larger bodies of water, including the coastline? Or why dumps and even industrial toxic liquid waste retention basins or settling tanks are often located along rivers? The answer goes beyond the historical advantage of access to water transportation: many industrial processes require large volumes of water, for example, as a coolant (creating thermal pollution, etc.) – or to conveniently wash away wastes. Waterways are monstrously misused as pressure relief valves! Fish kills, algal blooms, garbage cluttered and even burning rivers attest to the fact that the traditional maxim of wastewater engineers ("the solution to pollution is dilution") is a horrible misjudgment. In general, the major sources of all types of marine pollution are land-based, with the pollutants reaching the sea indirectly via waterways and the atmosphere. This is equally true for marine debris. Rivers are a major pathway for debris (whereby balloons, plastic bags, or fireworks, etc., can reach the sea via the atmosphere). Thus, cities and industries far inland ultimately pollute the marine ecosystem. Even dry river beds are relevant when used as illegal dumping sites: the first heavy rains in spring or flash floods then sweep this material into the sea (ever wonder why you often see trash clinging to higher-lying rocks and draped around vegetation along riverbanks?) (Fig. 1.5). This makes virtually all of us accomplices in the criminal offense of marine pollution and of debris on beaches – perpetrators at the scene of the crime, so to speak.

Recreational beach users are perhaps the most immediate sources of land-based beach pollution. Why? Let's take a quick look at the beach vacation framework. Every visit requires a basic set of items that are either brought from home or bought on site. Protection from the sun, wind, and hot sand

Fig. 1.5 Garbage dumped into streams, dry stream beds, or canals is swept directly into the sea by the next good rainfall. Atlantic, Morocco

Fig. 1.6 Instant beach litter. How do you rate the chances of everything being taken back home in the evening? Pacific, California

requires towels, sunscreen lotions, sunglasses, hats, and special footwear. No trip to the beach would be complete without food and drink to slate hunger and thirst. Beach games, swimming, and other water sports require dozens of additional items from balls to masks and snorkels. Even just relaxing with a book or with music is much more pleasant under a beach umbrella on a towel or blanket with a snack and a cool drink (Fig. 1.6). Of course, any true equipment freak can expand this list indefinitely. Finally, as every parent can confirm, taking small children to the beach is akin to outfitting an expedition, easily doubling or tripling the amount of required gear. Children also considerably raise the likelihood of articles being lost. Sandy beaches are the closest thing to cosmic "black holes" for litter right here on Earth. Once dropped, any smaller item that is not immediately retrieved is likely to be claimed by the sand. Moreover, in the black hole analogy, the presence of litter anywhere reduces the inhibition threshold of others to discard waste, and deposited garbage always attracts a chain reaction of others adding their refuse to the pile. Serious retrieval efforts are usually confined to money, watches, and jewelry, not to sticky ice cream wrappers, sand-coated food remains, and the like. The first line of defense is to bring along a bucket or trash bag and make sure that your trash actually ends up there (and is taken home; Fig. 1.7).

Fig. 1.7 The road to marine debris is paved with good intentions. Putting the trash into a bag is not enough. The wind, birds, dogs, and – depending on where you live – more ornery wildlife will soon strew its contents far and wide. Mediterranean, Turkey

Fig. 1.8 Where does "land-based" marine debris come from? One source is woefully undersized, inadequately designed, and infrequently emptied beachside garbage containers. Mediterranean, Turkey

Trash containers at the beach are the second important line of defense. You'd think that, after generations, seaside communities would get the trash bins right. Somehow, however, they almost never fulfill all the criteria below (Fig. 1.8):

Fig. 1.9 Garbage heaps or dumps within eyesight of the water – only a gust of wind away from becoming marine debris. Mediterranean, Turkey

- Solid- or fine-mesh wall construction along with lid to prevent windblown losses, leakage, and loss of smaller items (pull tabs, bottle caps, cigarette butts) and to thwart foraging seabirds, dogs, or other wildlife.
- Fire-resistant plastic construction rather than metal (which rusts rapidly in the salty air).
- Proper dimensioning to accommodate peak use. Beachgoers should never be confronted with the option of putting garbage next to a filled container or at the back-beach upon discovering that no bins are available (Fig. 1.9).
- Heavy enough or firmly anchored to withstand the strong winds typical of shore environments.
- Strategic placement, i.e., at every entry/exit to the beach, at sufficient intervals, and far back enough to avoid highest tides and wave wash.

Even when all these points are met, the final and crucial one:

- Frequent collection

is often neglected. Even the best containers cannot prevent beach pollution if they are not emptied regularly (and in between scheduled times if and when the need arises during the peak season).

Of course, there is no lack of premeditated and deliberate, large-scale and criminal waste disposal directly at sea or over seaside cliffs. In many more places than you might imagine, however, it is perfectly legal to dump everything from construction material to municipal and other wastes at sea. This activity is theoretically governed by the 1972 London Dumping Convention, but as landfills overflow and new sites become more difficult to find (not to mention smaller island states that have no land available for dumps), ocean disposal is an alternative that won't go away, even for nuclear wastes.

Ocean-Based Sources

The second major source of marine debris is ocean-based. Such "offshore" sources are a more direct type of marine pollution, although in the case of beach litter, the items may have traveled great distances before being washed ashore. The villains here include recreational boating and cruise ships, commercial fisheries, merchant shipping, research vessels, military operations, as well as the offshore oil and gas industry. "Ocean-based" is somewhat misleading: the culprit is of course human activities at sea rather than the ocean itself.

Recreational boating and fishing are important sources. Millions of pleasure craft ply our waters, and the number has increased dramatically in recent decades. The clanking forests of masts at every marina attest to this trend. Each boat outing produces wastes. Some of this stems from the kitchen (galley), including food wastes and the full range of household garbage such as trash bags, plastic milk and water containers, beer and soft drink cans, egg cartons, meat trays, and other food packaging. The main recreational fishing wastes include monofilament fishing line, plastic fishing floats along with hooks and lures. Yacht designers are chiefly concerned with safety, handling, and speed, not necessarily waste retention. Toilet facilities (the "head") on board are typically spartan – if you stand up and turn around, you rub your butt on three walls and stub your toe – and the waste products are more often than not purged directly into the sea. Other wastes are also often thrown overboard with bravado. Today's environmentally aware mariner collects, separates, and stores – taking advantage of modern dockside services to dispose of garbage and to pump out bilge water- and waste-holding tanks. Unfortunately, harbors themselves are major producers of marine debris (Fig. 1.10).

Commercial Fisheries

The world's commercial fishing fleets consist of hundreds of thousands of vessels, ranging from the more traditional ("artisanal") quaint wooden boats to less picturesque, industrial hulks. Fishing vessels basically have the same problems as the pleasure craft above: cramped space and priorities other than waste collection. The temptation is strong to discard irreparable gear and other wastes at sea

Fig. 1.10 Boatyards, harbors, and docks are often marine debris hotspots. Atlantic, Morocco

rather than storing them for later disposal on land. Fishermen are often said to have one great advantage over farmers: they can reap without having to sow. Factoring marine debris into the equation, however, shows that besides taking out lots of fish, fishermen put in plenty of wastes. This includes nets, ropes and lines, traps, buoys, fish boxes, buckets and other containers, salt and ice bags, light bulbs and light sticks, rubber gloves, and many other characteristic items beyond normal galley wastes (Fig. 1.11). Unfortunately, the items discarded by fishermen are very often those that are most dangerous to marine life. Trawl nets or gill nets are designed to catch fish; when lost or discarded, such "ghost nets" continue their senseless harvest, even as fragments. Traps designed to capture crabs or lobsters continue to do so over and over again even when lost on the seafloor. Longlines with hooks continue to fish for their targeted prey and other unintended sea life even if they have broken loose and are never again retrieved.

In fairness, many of the losses in commercial fisheries are unintentional or accidental. Expensive lines may be lost at sea in storms; wear and tear as well as operator error or misjudgment can cause gear to break loose. In shallower waters, nets must occasionally be cut after becoming snagged on the bottom (sometimes on larger "marine debris" such as shipwrecks). Boats illegally fishing in foreign waters are known to cut their nets when approached by the authorities in order to avoid heavy fines. Whether accidental or intentional, fishery-related items litter beaches worldwide. The specific items typically reflect regional practices. A quick trip to the nearest harbor and a look to the gear stowed on fishing boats will give the marine debris detective a good idea of what the beach may have in store.

Fig. 1.11 The world's fisheries are the major "ocean-based" producers of derelict buoys, lines, nets, and other "mega-litter." Pacific, South Korea

The tide may slowly be changing. Why? Fishermen are the first to be confronted with marine debris, both floating on the surface, suspended in the water, and lying on the bottom. All involve costs, lost time, and hardship: propellers and propeller shafts may become entangled, engine cooling intakes clogged, and fishing nets filled with more debris than fishes. All can reduce a ship's maneuverability – always a dangerous proposition on the high seas. One idea to alleviate this situation is to recruit fishers: voluntary "derelict net recycling" programs are in effect in some places (Hawaii), and, elsewhere, fishers were being paid to bring derelict gear and other debris they retrieve from the sea back for proper disposal at harbor [18]. Again, money talks! Finally, strategies to mark nets and lines at regular intervals will one day enable most derelict equipment to be traced back – hopefully also an impetus to reduce "losses."

Commercial Shipping/Merchant Vessels

Sea trade once defined nations and continues to be the cheapest way to transport goods around the world. It is also a major source of vessel-related marine debris. The sheer number of boats is part of the explanation. And they are always on the go: a day in port is a lost day and just as expensive as sailing. Some of the wastes produced here are dumped intentionally, but lost cargo and the material used to package and brace cargo account for much of the accidental debris. In rough seas, cargo (often entire shipping containers) may be lost when ships are damaged, and other equipment stored on deck may be washed overboard. Hazardous materials are sometimes intentionally stored on deck so that they can be more easily jettisoned to prevent damage to or contamination of the boat in emergency situations. Finally, cargo (and fuel) is often released when the ships themselves break apart or sink in heavy storms. The world merchant fleet losses amount to several hundred larger ships per year.

Cruise Lines

Cruise lines have repeatedly been exposed as producers of marine debris. Although their ships represent less than 1% of the global merchant fleet, they may be responsible for 25% of all waste generated by merchant vessels [19]. Such "love boats" have much more space than most other vessels, but some have cut corners by dumping rubbish into the sea. Each passenger produces an estimated 3.5 kg (that's almost 8 pounds) of trash a day, multiplied times let's say 4000–5000 guests on a larger ship (and they are getting larger every year) … that's a lot of potential marine debris! Typical items include plastic shampoo and conditioner bottles, plastic hand lotion bottles, balloons, as well as paper and foam cups. These are sometimes embossed with the name of the cruise line – a helpful piece of evidence for beach detectives and beach cleanup organizers. For example, in 1993, a cruise line was fined $ 500,000 for knowingly discharging plastic trash (20 rubbish bags) in US waters off the Florida Keys. The key evidence for conviction in this felony case was provided by two passengers who videotaped the incident. In 2016, a cruise line was caught systematically pumping oily wastes into the sea through a "magic pipe" for nearly a decade [20]. The fine: a whopping $ 40,000,000! When fines talk, people listen. A crew member was the whistle blower. Concerned and informed individual citizens can make a difference!

The Military

The military is a highly unpleasant source of marine debris because its operations are subject to less open oversight and because their wastes – and losses – are sometimes considerably more dastardly than domestic waste. A case in point is the practice of dumping munitions into the sea. New generations of weapons mean rapid obsolescence, and surplus materiel cannot always be sold to armies and insurgents elsewhere. Any yacht owner who has inspected his or her sea charts in search of an anchoring site can attest to the extent of munition dumping grounds, firing practice ranges, and other off-limit military areas. Much of the dumping takes place close to shore, and the materiel can be shifted or brought to the surface by fishing nets and dredging operations. Storms and waves can deposit such material on the beach. 1.6 million tons of conventional munitions and 200,000 tons of chemical warfare agents were dumped in the North and Baltic Seas alone after World War II in a program titled "authorized disposal of hazardous ammunition by dumping at sea" [21]. About 80 conventional and chemical munition dump sites (including WW I) are known in shallow northern European seas. Human injuries on beaches are reported (2-4 cases per year), particularly from chemical warfare agents such as white phosphorus released from incendiary bombs or propelling charge rods. The former resembles amber and, if collected, can self-ignite (for example in your pants pocket) at 1300 °C, the latter ignite in a long jet of flame. Other material such as toxic explosives from mines and torpedoes are almost indistinguishable from stones. All this material has gotten uglier and more difficult to recognize after half century or more underwater. This has led to numerous beach closures.

Perhaps the crassest and deadliest of all marine debris is land mines. According to a United Nations estimate, more than 100 million land mines have been sown in 62 countries across the globe. One hundred and fifty civilians are maimed or killed each week, prompting various efforts and celebrity support. Some of these land mines lie buried on beaches in order to hinder invasion from the sea. Planting them is easy, clearing them a horror. Shifting sands and corrosion due to saltwater compound the problem on beaches (see chapter "Metal").

Military vessels – of which thousands exist – have many of the same problems as merchant vessels and commercial fishing boats: cramped quarters, long periods between ports of call, priorities other than waste retention. Nevertheless, a new level of awareness is setting in. Some improvements are strongly self-interest: adversaries were long able to follow the progress of foreign fleets, including submarines, by following their "trash trail." In another turn of events, floating dark-colored plastic bags became a major headache for warships in a recent conflict: they resembled floating mines. The debris reduction strategies include waste pulping (grinding or pulverizing), waste compactors, plastic processors to heat and compress plastic wastes to reduce their volumes, or incineration. Some of these help the wastes to be stacked and stored for later disposal on land; others merely make trash negatively buoyant, i.e., sinkable, or turn solid debris (when burned) into air pollution.

Oil Industry

The oil industry is not only a source of oil pollution but also contributes to marine debris, for example, tar balls, which fit the UNEP definition above. Not only do normal tanker operations put more oil into the sea than catastrophic spills, they also produce the same type of onboard wastes as cargo vessels. Moreover, offshore oil and gas platforms lose deck light bulbs, hard hats, buckets, plastic sheeting, diesel oil and air filters, along with technical items such as seismic markers and drilling pipe thread protectors. Multiply this times the estimated 25,000 platforms currently in operation! Most of these are concentrated in shallow coastal waters close to the shore, meaning that adjoining beaches tend to be most heavily affected. The problem is often most severe in enclosed or semi-enclosed shallow seas. A case in point is the Gulf of Mexico which, as we know, has a slew of other pollution-related problems such as oil platform blowouts and the world's largest "dead zone" due to eutrophication-related oxygen deficiency.

International Treaties

Marine pollution is a global problem, and the key to reducing vessel-related dumping is therefore international laws. The problem can perhaps best be exemplified by the fact that, not long ago, more waste was being dumped into US waters by vessels each year than the weight of the entire annual fish catch taken by the US fleet. A legal milestone in this direction is the International Convention for the Prevention of Pollution from Ships, better known as MARPOL (which stands for MARine POLlution), formulated by the International Maritime Organization (IMO). Additional regional laws are also in effect, in the USA, for example, the Marine Plastic Pollution Research and Control Act of 1987 (MPPRCA). Importantly, MARPOL *regulates* dumping but does not *prohibit* it. Thus, with the exception of plastic, most garbage can be dumped beyond the 3 mile limit from the coast if it is ground up small enough. And the various Annexes of the Convention are known to include phrases such as "to the extent practicable," i.e., the regulations should not endanger crew health or ship safety, create unacceptable nuisance conditions, or compromise combat readiness (for military vessels to the extent that they need to comply with MARPOL). The law does not apply to cases when garbage escapes due to damage to a ship. And it is, like most international agreements, binding only for those countries that have ratified it.

Removing Marine Debris from Beaches

The final line of defense – after implementing all reduction strategies – is the regular removal of litter left on or washed ashore. This can be the task of bathing attendants, beachside concessioners, hotel personnel, or municipalities.

The latter typically resort to heavy machinery that "rakes" the top sand layer and variously conveys the collected items into a container or compacter (Fig. 1.4). This activity generally escapes the attention of vacationers because it is done at night, leaving beachgoers in the dark about the actual, often high accumulation rate of debris. Tractor tracks, orderly rows of furrows, unnaturally smoothened sand surfaces, or sharply juxtaposed clean and littered swaths of beach provide clues to beach debris sleuths.

Large-scale mechanical removal has several drawbacks. Only established vacation resorts can meet the high purchasing and operating costs. Not all shorelines can be serviced: pebbles, cobbles, or rocky outcrops can clog or damage equipment. Even on pure sandy beaches, cleaning is restricted to areas between the high tide mark and vegetated or dune areas. Heavy machinery also cannot reach litter hotspots such as under boardwalks. Mechanical collection is also selective: cans, bottles, drinking cups, and the like are removed; cigarette butts (by far the most common items), splintered glass, and all smaller items are missed. Unwieldy nets, barrels, pallets, tires, etc. must be collected separately. Moreover, collection is indiscriminate: man-made items are not differentiated from natural objects. Natural seaweed and driftwood accumulations, however, serve as a home for a wide array of small organisms. Sea shells provide homes to hermit crabs and, as they become increasingly broken and abraded, ultimately contribute to sand formation. All are part of the beach ecosystem and should remain in place. Lastly, sandy beaches, as a natural ecosystem, are a habitat for many burrowing animals such as ghost crabs. And don't forget that endangered sea turtles lay their eggs here. None are adapted to having their habitat regularly "plowed" by heavy beach-cleaning equipment.

This calls for … no, not a superhero, but for you and me, ordinary beachgoers. Find out what you can do about marine debris [22]. Join local beach-cleaning activities or participate in one of the Ocean Conservancy's fantastic international coastal cleanups [23], or simply pick up and take home one item more than you brought to the beach. Become a beach detective, an activist, part of the solution. Good luck and good hunting!

1.4 What Can You and I Do?

The "positive" perspective on marine debris and beach litter – as opposed to most other types of marine pollution such as heavy metals or radioactivity – is that you and I can directly do something about it. We are all in some way involved in producing wastes, and we can all get involved in reducing them.

We constantly hear about how complex our world has become and how monolithic and immovable "the system" is. This tends to breed apathy and inaction. Importantly, however, every major trend and every large-scale public effort, action, or grassroots movement has initially been triggered by an individual person – someone who first recognized a problem or discovered a solution, someone who was dedicated and felt strongly about an issue, someone with courage and conviction. You can be such an actor on the marine debris stage!

You can make a difference on many levels. As a *beachgoer*, take home every item you brought to the beach. Sounds simple, but many psychological and sociological factors – and a good dose of plain laziness – help explain why this is easier said than done. You can start off by making sure to bring only the things you really need and that those items lend themselves to being collected and properly disposed of in waste containers at the beach or taken home. Unpack newly purchased beach items at home rather than on the beach. Remove food from its original packaging at home, and take only the amount you need in washable, reusable containers. You might also consider taking the extra step of bagging and properly disposing of litter items washed ashore or left by others. It would already make a big difference if everyone picked up and properly disposed of just one item more than they brought to the beach. More generally, as a *consumer*, one way to approach the problem is in the framework of the so-called "R"s [24]. Six "R"s are distinguished here: rethink, refuse, reduce, reuse, repair, and recycle. Most folks would spontaneously place their bets on recycling, but it is not ultimate "R." It is not applicable to all materials and may merely postpone the final deposition. Look at the other "R"s (and an important "U") to help avoid waste in the first place.

Rethink Do I really need this product at all? Can I make do with the model I already have? Does it have to be made of materials that are environmentally unfriendly, that will take ages to decompose, that are hazardous or toxic, or that are known to end up on beaches in large numbers?

Refuse Just say no! Do I really need to accept a product in an excessively packaged state? Can I purchase the same product or food item without packaging? Do I really need to have every item I buy put into a separate bag? Can't I simply refuse disposable products (plastic kitchenware, cheap lighters, single-use cameras, etc.) and pick the higher-quality equivalent?

Reduce Can I reduce or replace my consumption of environmentally unfriendly products with alternatives? Can I replace plastic bags with a fabric bag? What about reducing plastic water bottles and paper cups at the office with a water dispenser, reusable water bottle, or my own cup. Taking my own cup to the coffee-to-go counter would in itself prevent the discarding of millions of single-use cups every day? Or using a personal set of chopsticks instead of throwing away the disposable ones after each restaurant or take-out meal? Or bringing my own eating utensils to the beach instead of buying one-way/single-use plastic cutlery? Or bucking the trend to miniaturized food products offered in cute, convenient, individually wrapped, bite-sized portions that use considerably more packaging?

Some of these ideas are included in the US Environmental Protection Agency (EPA) tips on how to "unpackage your life" and reduce some of the more common marine debris items [25]:

- Use no or bring your own (fabric) bag.
- Carry a reusable water bottle.
- Pack a waste-free lunch.
- Bring your own cup.
- Slow down and dine in.
- Say no to straws.
- Avoid heavily packaged foods.
- Bring your own container and utensils.

Reuse Why immediately throw away the new paper or plastic bags when returning home from shopping? Reuse them for the next spree or for the garbage. Use thin supermarket produce bags again and again. Use empty containers for storing something else at home or in the garage.

Repair Do I really have to trash a product because some small component is damaged? You'll probably be surprised at the number of local businesses that have cropped up to repair almost anything.

Recycle Most products and their packaging are actually valuable resources. When I do finally discard an item, is there a special bin I can put it? Properly separated and sorted, many waste categories can be recycled, reducing the amount of energy and natural resources needed to produce new items. All glass bottles, for example, can be recycled indefinitely. This also reduces the amount of material dumped into landfills and helps keep wastes out of our waterways.

Of course, there are a few other "R"s that come to mind. One might be *Re-invent*: this involves writing to companies and asking them to redesign their products, for example, to make them out of more ocean-friendly materials or to at least reduce packaging (see below). Finally, the "R" most relevant in light of the present guide is *Remove* it!

These six (or more) "R"s are wonderfully complemented by a "U" strategy:

Upcycle Many old products can be used "as is" or be variously reworked to make completely new items. Anything goes! Glass fragments, bottles and their caps, tin cans, etc. are the raw materials for untold arts-and-crafts projects; wooden pallets can be used to make high-end, "shabby-chic" furniture. You'll be surprised at the endless options: enter almost any item, and "upcycle," into your search engine for many clever and unexpected suggestions.

Of course, reducing marine debris – and wastes in general – requires efforts at higher levels too. The first prerequisite is a general degree of public awareness about environmental issues. Then, government authorities must provide the necessary legal frameworks and physical infrastructures (separate waste bins, a garbage collection system, proper waste treatment and disposal sites, fines). Finally, business and industry need to show insight and commitment. They must produce materials and products that can be recycled; technologies must be developed that enable profitable further use; the composition of products or packaging must be clearly labeled and the symbols recognizable to consumers. Worldwide, systems are in place that show waste recycling to be a profitable business. Rather than putting a burden on society, implementing such systems creates what politicians have been promising all along: jobs, jobs, jobs.

Sound like mere hollow words or wishful thinking? No. Again, you as an individual are not powerless and can actually help get things moving. Contacting a beachside proprietor whose packaging litters adjoining beaches can work wonders to help reduce trash at the source. But you can go another step further. After you have identified a product that contributes inordinately or avoidably to hazardous or unsavory marine litter, you can write letters – oops, I mean emails and twitters – telling companies you are concerned and asking them to consider reducing the respective packaging or to initiate a redesign using new, more ocean-friendly materials. Not happy with the response? Then take it to the highest level: contact your local, regional, or national politicians and appropriate government ministries. Inform the appropriate environmental organizations. Let your friends and family know what you're doing and get them involved. Was there ever a better reason to start a shit-storm? Marine debris is an issue recognized and being pursued by a wide spectrum of authorities and agencies, and you may well find an open ear or be able to join in a targeted campaign. And don't forget, most beaches around the world are public property – they belong to you!

Beach Cleanups

Finally, let's return to the beach itself. Beyond leaving nothing behind after visiting the beach, you may wish to get more involved personally. How about seeing if there is an "adopt-a-beach" program in effect in your area? You could actively help keep "your" shoreline healthy, safe, and beautiful – and it certainly sounds a bit sexier than the usual "adopt-a-highway" scheme. Your group will probably be asked to clean the beach a few times a year.

The major event of the year, however, is the International Coastal Cleanup, which takes place in over 100 countries worldwide and in most US states (including many without seashores because the event also encompasses lake shores and riverbanks). This is the world's largest volunteer event for the ocean (more than 12 million persons since 1985!). In recent years, more than 800,000 people have participated annually. This event usually takes place on a single day in mid-September. It is organized by Ocean Conservancy, which has also produced detailed reports over the years about what and how much was found where. Check out their website (www.oceanconservancy.org) to find and sign up for the cleanup closest to you.

Of course, you can also start your own waterway cleanup, by yourself or with friends and relatives, with or without the support of an organization.

What You Might Need for Your Beach Cleanup Effort

- A hat and ample sunscreen
- Appropriate footwear (i.e., no flip-flops on rocky coasts)
- Enough water to drink
- A pair of gloves and/or a collecting gripper stick
- A simple first aid kit
- A reusable container such as a bucket to collect the items in (or, alternatively, a trash bag)
- A hard-shelled container with a tight lid to deposit sharp or otherwise hazardous items (fish hooks, syringes, etc.)

Accurate information on the types and amounts of debris can be used to help solve the problem over the long term. By evaluating the data from organized cleanups, for example, environmental groups can confront the worst offenders and bring about change. If you want your data to be officially recorded, use a cleanup data form (see Ocean Conservancy above), and you can optionally weigh the collected material with a hand scale and take photographs of unusual items.

What You *Shouldn't* Do

- Open any closed containers, even if they appear to be empty (see hazard symbols in Fig. 1.12).
- Attempt to pull large items such as tires out of the sand or maneuver heavy items such as oil barrels alone.

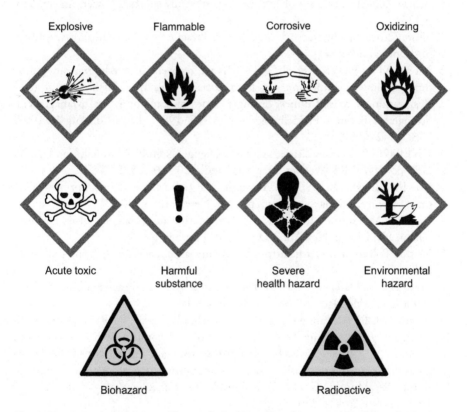

Fig. 1.12 Hazard pictograms. The symbols themselves have remained largely the same over time. The "frames" may differ (triangles, squares; new: diamond), as can the overall color (yellow, orange) or frame color (old: black; new: red). Marine debris labeled with any such symbol(s) should be handled only with a prior expert (beach cleanup coordinator) go-ahead.

- Attempt to remove unusual accumulations of potentially hazardous materials such as hospital wastes.
- Touch or move dead or injured marine wildlife.
- Remove natural biological items such as seaweeds, driftwood, jellyfish, and coral and shell fragments.

As a modern beachcomber, you may wish to consider a variation of the Hippocratic Oath taken by physicians: do more good than harm. May this book be a step forward in helping you become part of the solution.

References

1. Brown AC, McLachlan A (1990) Ecology of sandy shores. Elsevier, Amsterdam, p 328
2. Shore & Beach, American Shore & Beach Preservation Association. www.asbpa.org/shore-and-beach
3. Acosta ATR, Jucker T et al (2013) Passive recovery of Mediterranean coastal dunes following limitations to human trampling. In: Martínez et al (eds) Restoration of coastal dunes, Springer Series on Environmental Management. Springer, Berlin Heidelberg, pp 187–198. https://doi.org/10.1007/978-3-642-33445-0_12
4. UNEP (2009) Marine Litter: A global challenge. Nairobi; Marine Litter. UNEP/GPA & UNEP/RS. Wedocs.unep.org/handle/20.500.11822/7787
5. Coe JM, Rogers D (1996) In: Alexander DE (ed) Marine debris, Springer series on environmental management. Springer, New York, p 430
6. Bergman M, Gutow M, Klages M (2015) Marine anthropogenic litter. Springer, Cham, Heidelberg, New York, Dordrecht, London, p 447
7. Marine Pollution Bulletin. https://www.journals.elsevier.com/marine-pollution-bulletin/
8. Trewhella S, Hatcher J (2015) The essential guide to beachcombing and the strandline. Wild Nature Press, Plymouth, p 304
9. Timrott J (2015) Strandgut aus Plastik. Wachholtz Verlag, Kiel/Hamburg, p 112
10. Change.org. Accept the Trash Isles as an official country & help protect our oceans. https://www.change.org/p/un-secretary-general-ant%C3%B3nio-guterres-accept-the-trash-isles-as-an-official-country-help-protect-our-oceans
11. Rathje W, Murphy C (1992) Rubbish! The archeology of garbage. Harper Collins, New York, p 250
12. Parker L (2017) How an uninhabited island got the world's highest density of trash. https://news.nationalgeographic.com/2017/05/henderson-island-pitcairn-trash-plastic-pollution/

13. Gregory MR (2009) Environmental implications of plastic debris in marine settings – entanglement, ingestion, smothering, hangers-on, hitch-hiking and alien invasions. Phil Trans R Soc B 364:2013–2025. https://doi.org/10.1098/rstb.2008.0265
14. Mayell H (2002) Ocean litter gives alien species an easy ride. https://news.nationalgeographic.com/news/2002/04/0429_020429_marinedebris.html
15. Jang YC, Hong S, Lee J, Lee MJ, Shim WJ (2014) Estimation of lost tourism revenue in Geoje Island from 2011 marine debris pollution event in South Korea. Mar Pollut Bull 81(1):49–54. https://doi.org/10.1016/j.marpolbul.2014.02.021
16. Marine Debris Awareness Poster. https://web.whoi.edu/seagrant/outreach-education/marine-debris
17. Ocean Conservancy. https://oceanconservancy.org/wp-content/uploads/2017/04/OC-DataCards_volunteerFINAL_ENG-1.pdf
18. Cho D-O (2009) The incentive program for fishermen to collect marine debris in Korea. Mar Pollut Bull 58(3):415–417
19. Butt H (2007) The impact of cruise ship generated waste in home ports and ports of call: a study of Southampton. Mar Policy (5):591–598
20. Princess Cruise Lines to pay largest-ever criminal penalty for deliberate vessel pollution. The United States Dept. of Justice. https://www.justice.gov/opa/pr/princess-cruise-lines-pay-largest-ever-criminal-penalty-deliberate-vessel-pollution
21. Rudolph F (2015) Gefährliche Strandfunde. Wachholtz Murmann Publishers, Kiel/Hamburg, p 96
22. Trash-free waters. United States Environmental Protection Agency. https://www.epa.gov/trash-free-waters
23. Fighting for Trash Free Seas. Ocean Conservancy. https://www.oceanconservancy.org/trash-free-seas/international-coastal-cleanup
24. Walker K (2007) Recycle, reduce, reuse, rethink. Macmillan Education Australia, South Yarra, p 176
25. What you can do about marine debris. United States Environmental Protection Agency. https://www.epa.gov/trash-free-waters/what-you-can-do-about-marine-debris

2

Glass

2.1 Glass

What a fantastic material! Glass is transparent, gas-tight, acid-proof, tasteless and odorless, formable into any shape, stable, and does react with other materials. And the availability of its basic ingredient (sand) is virtually limitless. The ancient Egyptians already held glass in high esteem 3500 years ago. Today, mass production and untold uses in everyday life have made it commonplace, and commonplace tends to imply discardable. No wonder so much of it ends up as litter.

Maybe your science teacher once told you glass is actually a liquid? Forget that classical example of arcane school knowledge. It's certainly irrelevant and unhelpful in the case of marine debris – and becomes painfully obvious if you step on it. Glass on beaches comes in the form of bottles, jars and their fragments, and various types of light bulbs, along with glass ashtrays and the occasional windowpane. On some shorelines, you as a beachcomber might even find a glass fishing buoy or message-in-a-bottle, which means you have luckily bagged one of the "Big 10" desirable or at least tantalizing marine debris items (glass buoys, message-in-a-bottle, stranded watercraft, intact water sports equipment, money, watches, jewelry and other valuables, ship containers, dry drug packages, perfectly good apparel)! One noted message-in-a-bottle incident involved a rice wine bottle containing leaflets appealing for the release of a dissident: set adrift in China and found on Vancouver Island, Canada [1]. In a more recent case, a 132-year-old bottle – containing a message from a German oceanographic experiment in 1886 – was found on an Australian shore, making its finders instant international celebrities [2].

The previous Guinness World Record for the oldest message-in-a-bottle was 108 years.

Glass items are the most abundant nonplastic debris category (let's leave cigarette butts out of the equation for the moment) and are high up on the list of the top 10 or "dirty dozen" items collected during international beach cleanups. Glass is not only very abundant on beaches, it is definitely the most long-lived litter category: it withstands saltwater and sunlight much longer than metal or plastic. Glass is basically made by heating sand (and a few secret ingredients) until it melts and then pouring the molten mass into forms. Does this make it a harmless, environmentally friendly waste product? No. An "undetermined" degradation time, i.e., longer than human history, has been estimated for bottles in the environment [3]. Embedded in sand or washed into the dunes, glass can no doubt remain intact for centuries – a boon to bottle collectors but bane to beach strollers. Of course, the breakdown process is accelerated in the surf zone, especially on rocky shores, or when crunched underfoot, which brings us to the hazards-of-marine-debris issue, in this case the threat to human health. Stepping on glass has become such a problem that most signs regulating beach conduct put "no glass containers" near the top of their list of rules (Fig. 2.1).

Fig. 2.1 It's etched in stone (or wood in this case) – as a health hazard, glass is often a priority no-no in beach rules, but this addresses only one type and one source (beach-goers) of marine debris. Atlantic, USA

2.2 Glass Bottles and Fragments

Bottles come in an infinite variety of sizes (volumes) and forms, making the term "bottle-shaped" meaningless. Other distinguishing features include color, thickness, ornamentation, and type of opening, but experts recognize dozens of additional characteristics [4, 5]. Importantly, bottle design is brand identity. Companies take great pains in creating distinctive, instantly recognizable bottles for their contents, whether this be beverages, perfumes, or medicines. International markets and global advertising dictate uniform corporate designs for a specific product to help us consumers conveniently find and buy "our" brand worldwide. This makes bottle redesigns and face-lifts risky and rare, which is why companies taking the leap often need penetrating public relations and advertising fanfare ("our new bottle – 10% more content and now even easier to pour!"). Such stylistic inflexibility helps beachcombers to recognize most bottles and their contents. Of course, persistent sleuths can browse the aisles of local food and liquor stores to find the match. On fresh bottles, check for adhesive labels, raised (embossed) or other forms of lettering, and even caps: they typically have brand names and logos that can help you identify otherwise unintelligible foreign-language labels.

How do bottles end up on beaches? In beach litter lingo, the sources range from shoreline and recreational activities, to dumping activities, to ocean and waterway activities. Translation: it's you and me. Pick your excuse for leaving empty or half-full bottles behind on the beach:

- Too heavy.
- Too many.
- Too fragile.
- Too sticky.
- The wind blew away my plastic trash bag.
- Loose bottles are too dangerous/too noisy in the car.
- I lost the cap and didn't want to pollute the beach by pouring out the remaining icky beverage.
- I buried it deep enough to avoid any problems.
- Someone will be happy to collect them for the deposit money.
- Hey, I pay my taxes to have this stuff collected.
- Officer, I drank it all and am too drunk to remember.

Then there are the boaters who throw them overboard. Excuses please:

- Let's see if we can throw them in so that they sink.
- Only an empty cooler can be refilled.
- They make great homes for marine organisms on the seafloor, don't they?
- Hey, glass is nontoxic.
- I wanted to write a message-in-a-bottle but forgot the message.

Whether discarded on the beach or washed up by the waves, bottles can persist virtually forever, especially if they end up buried or nestled somewhere in the dunes. This makes for long-term crime scenes. Of course, they can break in the surf or be crushed by vehicles, yielding "sea glass," a term which creates the unfortunate impression of being a legitimate component of the marine environment. Some refer to the worn and opaque "frosted" ones as "mermaid's tears." I agree: they are a crying shame! Fresh fragments (sharp-edged, clear) can be distinguished from older ones (rounded, frosted). Just like a newspaper scattered by the wind, one glass bottle can produce lots of fragments. The thickest and most reinforced parts (bottoms, necks, mouths) tend to remain intact longest. Some original products can be recognized even based on such fragments: look for unusual shapes, thickness, color, or remnants of embossed lettering.

Hazards: The fact that glass, like certain other categories of marine debris (wood, metal, etc.), is per se nontoxic doesn't make it any less hazardous to the environment or to human health. To begin with, bottles – capped or not – can float for long times and cover great distances. They are overgrown by marine organisms that can be rafted on what might be termed a "floating hotel" to faraway habitats. This hitch-hiking brings exotic species to places they normally would never reach and where they shouldn't be. Such "aliens" can be as evil as their name implies: they sometimes outcompete and displace native species, disturbing ecosystem balance and posing a threat to biodiversity. Finally, glass fragments can function as magnifying glasses and cause fires in strong sunlight.

As far as human health is concerned, most discarded containers, including bottles, tend to retain at least a bit of their original contents. "Good to the last drop" is advertising hype that rarely meshes with reality. Use caution when handling bottles potentially containing harmful remnant fluids. Check the labels for hazard symbols. Some, like a skull-and-bones for "poisonous," can be understood intuitively, others not (Fig. 1.12). The far more common threat, however, is accidentally stepping on bottles and glass shards. While entire bottles can be recovered by mechanized beach-cleaning equipment on larger public beaches, smaller pieces may pass through the meshes. Sharp glass fragments translate into foot injuries and are a compelling reason to forgo the pleasure of walking barefoot. Wear adequate footwear to prevent foot injuries during beach cleanups. Never throw bottles into barbecue fires: they tend to burst and complicate removal. A fresh, sharp fragment is rarely found alone: if you see one, take a closer look around you. Chances are good that the original bottle or what's left of it is lurking somewhere nearby. Such breakdown processes, whether intentional or not, are one explanation for the "law of marine debris aggregation" – the fact that if you see a particular item, you are likely to find more of the same at that particular spot. This, by the way, also holds true for intact bottles: or have you ever heard of a booze party on the beach where some people leave their bottles behind but others don't? Well-rounded larger pieces should also be removed because the can re-break, starting the cycle anew. Those were the don'ts. What about the do's? Take a few more bottles back home or to the beachside trash bin than you brought with you!

Alternatives: Rethink, refuse, reduce, reuse, recycle, and upcycle – that's 5 Rs and a "U". Each of these conservation mantras applies to glass. Rethink: can I put my beverage in a thermos bottle that I'll be sure to take back home? Refuse: can I purchase my drink in a returnable bottle instead of a throwaway one? Reduce: can I buy one larger bottle instead of numerous "personalized portion" bottles? Reuse: can I refill the bottle or use it to store any other liquids in the garage or basement, to water the plants or for something else? Recycle: this means making new bottles from old bottle glass. Glass is perhaps the most recyclable of all materials. It can be melted down and formed into new products indefinitely, without changing its properties. All bottles can and should be recycled. There is no doubt a recycling bin for glass somewhere near you. An admirable approach is returnable and refillable bottles. While this does involve cleaning costs for the bottling industry, it still uses much less energy than melting sand or old glass anew to make fresh bottles. You can support this by not buying beverages in "no return, no deposit" bottles. Beverage container deposit legislation or "bottle bills" help guarantee that bottles are returned. In the USA, for example, the lowest percentages of bottles and associated items on the beach were found in states with such legislation and recycling programs.

The final item is upscaling. The Internet abounds with "101" do-it-yourself ways to recycle bottles and jars. One site boasts 735 glass bottles upcycled images [6]. And, finally, what about the glass pieces? Many folks "upcycle" them for arts-and-crafts projects. In fact, entire associations – complete with meetings and festivals – are devoted to upcycling sea glass [7]. Use them to fashion decorative patterns or sea creatures on cardboard or (drift)wood. Or fill a glass bowl with the abraded, frosted ones as a decorative centerpiece. Add water to keep their beachside glow. What a heyday for amateur jewelry makers too. Check out the Internet for the many other upscaling ideas for bottles and fragments!

Fig. 2.2 Glass bottles exemplify marine debris by combining many components – paper, glass, cork, plastic, and/or metal. Most start out their beach lives with labels, making life easy for beach detectives. Atlantic, Belgium

Fig. 2.3 Caps and plastic or metal rings survive longer than labels. Distinctive shapes help identify the original product: browse your local grocery or liquor store shelves and find the match. Mediterranean, Italy

Fig. 2.4 Small, cute, "personalized" portions mean more packaging and waste, less content, and higher profits. Don't play the game: rethink, refuse! Mediterranean, Turkey

Glass 37

Fig. 2.5 Glass bottles always make the list of top 10 items found during beach cleanups and are by far the most long-lived marine debris. On beaches, buried bottles can reappear unscathed decades or centuries later. Mediterranean, Italy

Fig. 2.6 You can fool some of the people some of the time. Pacific, USA

Fig. 2.7 and Fig. 2.8 Telltale aluminum cap and central rubber seal: medical vial! Most (but rarely all) the contents are extracted by hypodermic needle through the seal. Keep an eye out for other hospital paraphernalia that might also have washed ashore! Mediterranean, Turkey.
Inside view of medical vial cap with rubber stopper and glass remnants held firmly in place by aluminum ring. Mediterranean, Greece

Fig. 2.9 Welcome to the "law of marine debris aggregation"! You'll rarely find one bottle alone on a tourist beach, more typically several of the same or competing brands like this fresh aggregation of beer bottles. Mediterranean, Turkey

Fig. 2.10 After being abandoned by their patrons, Fred and Ethyl (slightly tipsy) spent the rest of eternity together watching glorious sunsets on the sea. Mediterranean, Turkey

Fig. 2.11 It may be more elegant not to drink straight from the bottle, but glasses are thinner, less well visible, and particularly likely to crunch under your bare feet. Mediterranean, Turkey

Fig. 2.12 Ouch! One bottle, many sharp fragments waiting to be stepped on. Cap still attached. Collect bottles while they are still intact. Mediterranean, Italy

Fig. 2.13 It doesn't take a beach detective to know that glass – like this beer bottle – breaks quicker on "harder" gravel or cobble beaches. Mediterranean, Turkey

Fig. 2.14 Warnings, pictograms, instructions, and list of ingredients – nothing about leaving glass bottles on the beach though. Mediterranean, Turkey

Fig. 2.15 Tube-shaped bottlenecks and thicker lips – often protected by caps and metal or plastic foil – remain intact and recognizable longest. Mediterranean, Italy

Fig. 2.16 Glass bottle lip protected almost indefinitely by thick band and robust cap. Mediterranean, Turkey

Fig. 2.17 Reinforced and indented base of bottle. This shape helps withstand pressure (and simplifies turning by hand in traditional champagne cellars). Mediterranean, Turkey

Fig. 2.18 Opaque, "frosted" glass shards with variously rounded (aged) edges. Check out the Internet for endless decorative uses. Mediterranean, Greece

Fig. 2.19 Burning debris on the beach is seldom a good idea: intact bottles often burst, pose an even greater hazard, and are more difficult to remove. Mediterranean, Turkey

Fig. 2.20 Pane means pain. Window glass is relatively rare beach debris (and, unfortunately, cannot be used by bottle glass recycling facilities). Can't you just hear it crunch underfoot? Mediterranean, Turkey

Fig. 2.21 Congratulations, you've just found one of the "Big 10" desirable marine debris items! Glass floats were designed for the nets of Japanese fishing vessels and originally made by glassblowers from recycled sake bottles, now replaced by aluminum, plastic, or styrofoam – a step backward in the recycle-reuse-rethink philosophy! Pacific, USA

Fig. 2.22 Don't drink and drive? Don't drink and drop! Mediterranean, Turkey

Fig. 2.23 Smoothly indented upper rim. The beach detective's conclusion? The remains of an ashtray! Adriatic, Slovenia

2.3 Light Bulbs

The conservation crowd associates one concept with light: light pollution. This is a major issue for a full range of wildlife, killing untold insects, leading birds astray, preventing sea turtle hatchlings from reaching the sea, and modifying behavior and disrupting activity patterns in general (not to mention human health and psychology). On the positive side, light pollution is among the easiest forms of pollution to eradicate ("just switch it off"!). The marine debris issue, however, involves the light sources themselves.

Incandescent light bulbs, fluorescent light tubes and lamps, halogen lamps, and energy-saving lamps – whatever your light philosophy – they shouldn't end up on the beach. They exemplify marine debris in that they float well and actually consist of many combined materials: the glass bulb or tube itself, a metal screw base or other metal fitting, metal filaments, plastic or ceramic insulation material, and an enclosed gas. And they pose a health hazard.

Lighting elements come in many shapes, sizes, colors, and types of fittings (screw base versus bayonet fitting). These can help beach detectives distinguish the type of bulb and its intended duty. Wattages and other information stamped onto the glass provide further clues.

And why do light bulbs land on beaches? They come from municipal waste, deck lights on oil and gas platforms, boats, lamps used in commercial fisheries, and lighting of beach restaurants and bars. None are made to last forever. In fact, light bulbs were early on at the center of the "planned obsolescence" debate after their life spans were "engineered" down in the 1920s [8]. Enormous numbers of light bulbs are used to illuminate the world's 10,000 oil and gas platforms at night. This makes some light bulbs "operational wastes" that are particularly abundant on shores of oil-rich seas such as the Gulf of Mexico. Coastal cleanups yield tens of thousands of light bulbs and fluorescent tubes each year: the 25-year summary by the Ocean Conservancy tallied 438,361 [9]. Powerful lamps are also used in commercial fisheries. On moonless nights, they help attract a wide range of species, especially squid, to the surface where they can be more easily caught. The illumination produced in such fishing operations is clearly visible on satellite photographs of the Earth taken at night.

Like most marine debris, lighting elements are hazardous to both humans and wildlife. The thin glass means they *will* break when you step on them. Not only are their knife-like edges not blunted by wave action or sand abrasion, but the projecting remains are often held firmly in place by the metal cap and will show no give under your bare feet.

No light bulb (or any other product for that matter) was ever designed with its fate in the oceans or impact as marine debris in mind. In the name of energy savings and reduced greenhouse gas emissions, many countries are phasing out incandescent bulbs. New generations of light bulbs, although more expensive than the traditional product, use less electricity, burn much longer, and could help reduce this form of glass debris. On the other hand, some contain toxic elements. Among them is mercury, which enters food chains and is one of the key human health concerns in fish products. The take home message: make sure you dispose of every used light bulb correctly.

Fig. 2.24 Light bulbs: typical marine debris – they float well, consist of several components, and get ugly fast in saltwater.

Fig. 2.25 The bayonet-type cap and large size point to duty at sea. Red Sea, Jordan

Fig. 2.26 and Fig. 2.27 Large size and unconventional cap point to fishing fleet lights (caliper opened to 1 cm). Note blue strapping band underneath – you'll seldom find a lone marine debris item. Pacific, Japan.
The metal caps often rust and break away first, exposing the glass stem and sharp internals. Caribbean, Cuba

Fig. 2.28 and Fig. 2.29 Barnacle encrustation reveals time spent at sea and even the floating orientation. Such hitch-hiking organisms can float thousands of miles, ending up as "alien" species in faraway habitats.
Barefoot hell: crushed and broken bulb, exposed stem with metal filament and cap. Mediterranean, Turkey

Fig. 2.30 and Fig. 2.31 Metal screw cap with glass stem and intact wires. Mediterranean, Turkey.
Final decomposition stage of detached lightbulb aluminum cap; the glass remnants are bound to be around somewhere though. Mediterranean, Turkey

Fig. 2.32 and Fig. 2.33 Fluorescent (neon) light bulbs float well too! Nicely bedded between plastic cup, corn husk, and washed-up seagrass. Mediterranean, Turkey.
The fitting on the right side has fallen out. Guaranteed to crunch nicely underfoot. Pacific, Japan

Fig. 2.34 and Fig. 2.35 Aluminum fitting (pin base and pins, insulation) and just enough glass to make it hurt. Red Sea, Jordan.
You have to be a good beach detective to recognize the final stage of a neon lamp: aluminum fitting, no pins, insulation, or glass. Mediterranean, Turkey

References

1. Ebbesmeyer C, Jr Ingrahem WJ et al (1993) Bottle appeal drifts across the Pacific. Earth Space Sci News (EOS) 74(16):193–194. https://doi.org/10.1029/93EO00165
2. Oldest message in a bottle found on Western Australia beach. http://www.bbc.com/news/world-australia-43299283
3. Marine debris is everyone's problem. https://web.whoi.edu/seagrant/wp-content/uploads/sites/24/2015/04/Marine-Debris-Poster_FINAL.pdf
4. Lindsey B (2018) Bottle Glossary. https://sha.org/bottle/glossary.htm
5. White JR (1978) Bottle nomenclature: a glossary of landmark terminology of the archeologist. Hist Archeol 12(1):58–67. https://link.springer.com/article/10.1007/BF033734400
6. Glass Bottles Upcycled. https//:www.pinterest.com/weezie64/glass-bottles-upcycled/
7. North American Sea Glass Association. https://seaglassassociation.org
8. Wong C (2012) Planned obsolescence: The light bulb conspiracy. http://melaniebourque.webs.com/history%20of%20lightbulb.pdf
9. Ocean Conservancy (2011) Tracking trash, 25 years of action for the ocean. https://issuu.com/oceanconservancy/docs/marine_debris_2011_report_oc

3

Metal, Vehicles, and Tires

3.1 Metal

Plastic floats and metal sinks, right? Wrong. Ten thousand tons of metal can ply the waters (ships), and 100 tons of metal can even fly (airplanes). Just as amazing, metal marine debris of all types can float under the right circumstances. And, of course, ends up littering our beaches.

Metals make up most of the Earth's weight and volume, and most of the 100 elements or so in the periodic table are metals. Nonetheless, they are not really part of biological life cycles, so we are not normally confronted with them in nature. Human ambition – faster, higher, and further – has meant unearthing enormous amounts of metals and creating ever better and specialized alloys. The discovery, processing, and use of metals have even defined eras in the cultural evolution of humankind ("The Bronze Age"). Some metals are toxic in tiny amounts. Mercury, for example, is one of the few forms of marine pollution that has actually killed many human beings [1]. On the marine debris front, however, we are interested in the more tangible metals. Metals have come to play such an important role in our daily lives that they are components in almost every item we use. They often literally surround us for hours every day (think automobiles).

How do metals land on beaches? Some are brought by tourists and left there. This includes everything from beverage and spray cans to barbecue grills. Aluminum foil and other picnic-related items are commonplace. Some folks apparently consider beaches to be good places to deposit refrigerators, car parts, or even entire vehicles, much as their ilk would do elsewhere in the landscape. Metal debris ranges from ugly to potentially lethal. The latter includes rusty gas cartridges (Fig. 3.44) and land mines. I'm kidding, right? No (Fig. 3.45). What are the statistics? Up to 200 million land mines scattered around the world, tens of thousands of civilian casualties (often decades later), wide swaths of no man's land in dozens of countries, international conventions [2] to ban them (not ratified by all, including you know who, and then applicable only to antipersonnel, but not to anti-vehicle mines), cover titles in major publications [3], celebrity advocates, and no end in sight. And no wonder some are also found along shorelines. Of course, some of the more modern ones are made of plastic to avoid metal detectors…

Metals can also wash ashore from afar. There are basically three possibilities: the metal can be part of another object that floats, the metal object can bob on the surface like a boat by displacing water, or air is trapped inside. The spectrum ranges from nails in wooden boards to beer cans, ship containers, and derelict boats. In one highly publicized case, a Harley-Davidson motorcycle that was swept into sea (along with millions of tons of other wreckage) in the Fukushima earthquake and tsunami disaster in 2011 washed ashore a year later in Canada. It floated 5000 kilometers in its parking container across the Pacific. Its owner was identified and the motorcycle sent to a place of honor in the company's museum in Milwaukee [4]. In a more recent case, metal airplane parts washed ashore on Reunion Island in 2015 and helped refine the (unsuccessful) search for an ill-fated Malaysian passenger plane that disappeared over the Indian Ocean a year earlier. There are also more pleasant potential discoveries of metal on beaches. This is why so many public beaches are scoured by folks with metal detectors. The potential prize: three of the Big 10 desirable beach litter items, namely coins, jewelry, watches, and, for the very lucky, historical gold coins from ship wrecks of past centuries.

The strength of metals makes them indispensable for modern life. It also means they remain in the environment for a long time. Many are also variously treated or painted to prolong their useful lives, and certain alloys made for use at sea are very hardy indeed. Nonetheless, while not biodegradable, even the most massive and most high-tech metals ultimately stand no chance in the marine environment. The catchword is salt. The sea is simply one of the most corrosive habitats imaginable. This means rust and more rust (or oxidation and more oxidation for the nitpickers). It also means ugly and hazardous. Some metals pose problems for wildlife. A case in point is the traditional lead-containing items that cause environmental damage in marshes and poison birds after decades of hunting with lead shot [5]: new fishing sinkers and weights as well as shotgun shell pellets had to be introduced. Any metal is a barefoot human health threat on our shorelines. This means that all metal items should be collected during beach cleanups. Wear gloves and use caution when picking up jagged, rusting pieces. Larger items such as oil drums or vehicle remains, often partially buried in the sand, may require machinery to be removed. During beach cleanups, the coordinators should be consulted before handling larger items and larger liquid containers (they are never empty to the last drop, and, simply put, any liquid remnants heated up by the sun for days or months are bound to be toxic even if they weren't so originally).

All metals, regardless of their condition, are valuable. This explains why scrap metal is bought and sold, why aluminum cans are scavenged out of garbage bins, why metals are increasingly being salvaged from electronic devices, and why the nefarious come up with the idea of stealing copper wiring and cables, even at the risk of causing train accidents and other catastrophes. The value of metal – or the major energy and money savings of melting existing metals rather than producing them anew – is a prime driver of the recycling philosophy. Join the movement and make sure none of your metal waste ever lands in a landfill – or on the beach – again. You'd also be surprised at the endless arts-and-crafts, DIY projects, and other upcycling ideas for metals – whether it be forks and spoons, old wrenches, barrels, the internals of washing machines, or even pull tabs [6]. Enter "upcycle metal" in your computer's search engine for more clever ideas!

Fig. 3.1 Aluminum cans: technological marvels, works of art, valuable raw material, ecological madness. No matter what your take on the above, always on the "top 10" list of the most common items collected during international beach cleanups. Caribbean, Cuba

Fig. 3.2 Burial in dry sand can prolong their life for decades (200 years according to one estimate [7]). A bit of lettering, part of a logo: enough to identify the evidence. Atlantic, USA

Fig. 3.3 The graphics on aluminum cans fade first, then the thinner walls deteriorate. The razor-sharp edges are one reason for wearing gloves during beach cleanups. Caribbean, Cuba

Fig. 3.4 Crushing or burning cans doesn't make them go away any faster. Note rusting metal lid with glass fragment on right. Nothing for bare feet! Mediterranean, Turkey

Fig. 3.5 Yes, metal floats. The encrusting organisms (barnacles, bryozoans) on this can point to time spent at sea. This can introduce "alien" species to faraway habitats. Atlantic, USA

Fig. 3.6 Some cans are made of two metals, an aluminum top and a steel body, which rusts first. Mediterranean, Turkey

Fig. 3.7 Aluminum also ultimately oxidizes (rusts) in the salty environment, leaving thin, sharp, jagged edges. Mediterranean, Turkey

Fig. 3.8 and Fig. 3.9 The thickest part, the pull tab, survives the longest. New models that do not separate from the lids had to be introduced because birds and other wildlife swallowed them (but they still ultimately detach). Mediterranean, Slovenia.
This aluminum model has a plastic pull-ring and plastic seal inside. It also provides yet another component on which proud companies can put their name and logo. Mediterranean, Turkey

Fig. 3.10 and Fig. 3.11 Caps and lids are always ranked high in beach cleanup "top 10" lists. You'll seldom find only a single one ("law of marine debris aggregation"). Logos help beach detectives reconstruct the party. Caribbean, Grenada.
A tough one for beach detectives: the plastic seal of a bottle cap, which survives longer than the metal. The "crinkled teeth" of the cap are mirrored in the seal. Red Sea, Jordan

Fig. 3.12 Drinking on the beach produces the full range of marine debris categories: metal, glass, paper, plastic, and typically cigarette butts… Mediterranean, Turkey

Fig. 3.13 …and other beverage paraphernalia like this insulating can holder to keep your drink cool. Atlantic, USA

Fig. 3.14 The "law of marine debris aggregation" applies to cans as well: inconsiderate beachgoers rarely leave only a single one behind. Mediterranean, Turkey

Fig. 3.15 Spray cans on beaches are often related to the tourist industry (insecticides) or various personal hygiene and skin care products. Caution: potentially explosive. Mediterranean, Turkey

Fig. 3.16 Boating and construction rely heavily on foam insulation in spray cans. Mediterranean, Turkey

Fig. 3.17 The graphics fade, the metal body rusts, and the plastic parts are almost as good as new. Caribbean, Cuba

Fig. 3.18 Rusty spray cans are a hazard for the barefooted – collect them all during beach cleanups. Mediterranean, Turkey

Fig. 3.19 The thickest parts survive the longest. Mediterranean, Turkey

Fig. 3.20 Even when spray cans rust away to nothing, the plastic internals ("feed pipe") simply won't fade away. Mediterranean, Turkey

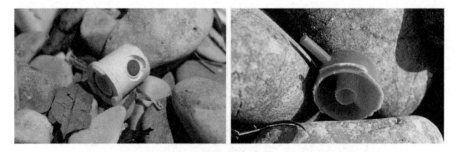

Fig. 3.21 and Fig. 3.22 Spray can nozzles are virtually indestructible. Mediterranean, Greece

Fig. 3.23 Canned food means tin cans, even on beaches. The labels provide all the information that beach detectives need. Atlantic, USA

Fig. 3.24 Labels are soon lost, rust sets in, and sharp lids await a soft passing foot. Red Sea, Jordan

Fig. 3.25 The older and rustier, the sharper! Mediterranean, Turkey

Fig. 3.26 and Fig. 3.27 Canisters like this are sturdily built and typically contain motor oil or cooking oil. Mediterranean, Turkey

Fig. 3.28 and Fig. 3.29 You can never get the last drop out of any container. Leaking residues mean extra caution. Never open a sealed canister and wear gloves during beach cleanups. Mediterranean, Turkey.
Sturdy or not, metal canisters are ultimately no match for salt water and surf. The plastic handle and spout remain unscathed though… Mediterranean, Turkey

Fig. 3.30 and Fig. 3.31 …and live to tell beach sleuths the tale long after the canister itself has disintegrated. Mediterranean, Turkey.
The final decomposition stages of canisters are an accident waiting to happen. Mediterranean, Turkey

Fig. 3.32 This "tin can" is a 200 liter oil barrel complete with labels, addresses, and telephone numbers. Call and complain! Caution: they will never be completely empty. Atlantic, Scotland

Fig. 3.33 Most barrels lack identification, contain harmful remnants, and are partially buried. Use caution and watch your back during cleanups! Atlantic, Scotland

Fig. 3.34 and Fig. 3.35 Other metal objects like this refrigerator can contain hazardous coolants and many plastic parts and require good backs to remove. Mediterranean, Turkey.
The enamel coating on major metal household appliances ("whiteware") only briefly delays rusting on beaches. Atlantic, Scotland

Fig. 3.36 Another reason why they don't want you to take shopping carts out of the store: they're worried about the marine debris problem. 149 of them were collected during a recent international coastal cleanup. Caribbean, Grenada

Fig. 3.37 and Fig. 3.38 Unfortunately, rust soon renders even potentially useful beach finds like these pliers worthless. Mediterranean, Turkey.
This teapot was probably once used to cook tea on the beach – a favorite pastime on many Mediterranean beaches. Turkey

Fig. 3.39 and Fig. 3.40 What's the beach detective's take on this left-behind grill? Too greasy? Still too hot to take back home? Or did a flimsy leg buckle? 26 were collected in a recent international coastal cleanup. Pacific, USA.
Beyond single-use aluminum grills (sold complete with charcoal), this type is most commonly left behind. Mediterranean, Turkey

Fig. 3.41 Hazardous metals on beaches include fish hooks, which are equally as good at hooking your foot as catching fish. Check out the "fishing gear" chapter. Mediterranean, Turkey

Fig. 3.42 A measuring band: up to 5 meters of rusty, sharp-edged metal waiting to snag the unwary. Mediterranean, Turkey

Fig. 3.43 "Box spring" seats and mattresses also pack lots of nasty metal. Mediterranean, Turkey

Fig. 3.44 The second most dangerous metal debris is rusty gas cartridges. These are potential bombs. Handle with care! Mediterranean, Ibiza

Fig. 3.45 And this is a bomb! By far the most lethal beach litter I have ever encountered: a land mine remaining after a Middle East conflict. A scientist and several dogs were killed during my stay. Red Sea, Gulf of Suez, Egypt

Fig. 3.46 Shipwrecks and grounded boats are potentially one of the "Big 10" desirable marine debris items, but this one is a job for the navy or coast guard, not beach cleanup volunteers. Of course there is always an interesting background story for beach detectives! Pacific Ocean, California

Fig. 3.47 Waiting a few decades is also a common strategy for larger derelict ships but yields highly unstable, rusting hulks. Beachgoers and swimmers – keep your distance! Indian Ocean, Maldives

Metal, Vehicles, and Tires   67

Fig. 3.48 Abandoned research stations and whaling centers leave metal beach litter of an entirely different dimension behind. Antarctic, Deception Island

Fig. 3.49 Sooner or later, experienced beachcombers will find everything including the kitchen sink, which is typically made of metal. Atlantic, Madeira

3.2 Vehicles

Vehicles and beaches. You'd think that getting away from the din of the automobiles, trucks, and motorcycles that overrun our cities and towns would be a prime reason to escape to the beach. Only the sound of the gently lapping waves and an occasional sea gull… Unfortunately, that's a bit naive these days. Most folks aren't big walkers, and you can't carry all the equipment you take to the beach very far anyway. Most visitors would just as soon like drive up to the very spot they lay their towel on. Besides, everyone likes to keep an eye on their precious cars rather than have them parked somewhere out of sight. This helps explain why you'll see cars on the sand or as close to the beach as possible when barriers exist.

This also means you'll find tire tracks crisscrossing many beaches, both on those where driving is allowed and those where it is not. Sometimes it will be fishers driving to a remote spot on a long beach or beach patrols in marked vehicles or riding quad bikes. Other times it will be regularly operating beach cleaning machinery (Fig. 1.4) or heavy equipment to get the beach "in proper shape" during the preseason or after the sun sets and most folks have gone home. One important thing needs to be said about sand, saltwater, and vehicles: they simply don't mix. The average car doesn't get very far at all on soft sand. But everyone has to make their own mistakes, and millions of drivers have to discover this for themselves. Those in the know are equipped with wide-tired 4-wheel-drive SUVs and pickup trucks. And many are out for some rambunctiousness, to "test" their vehicles and to validate having spent so much more money on a gas-guzzling all-terrain-type vehicle. The result is more often than not mayhem on vulnerable beach ecosystems. Driving on beaches and dunes destroys fragile vegetation and promotes erosion. On sea turtle nesting beaches, it also compacts the sand, hindering baby sea turtles from emerging from their nests, and it creates deep ruts that trap and misorient the hatchlings on their dash to the sea.

What does this have to do with marine debris? Well, a car that's stuck in the sand can already be thought of as beach litter, albeit an expensive one that won't be left behind long. Not counting the wrecks that are purposefully abandoned in beach landscapes. Importantly, vehicles are not self-contained devices: they tend to lose parts like hubcaps, mud flaps, and license plates when up against beach conditions. They also leak oil whether driving or standing (just check out the oil blotches on any parking area when the cars are gone over the weekend). And folks who throw things out the window on the highway are likely to do the same on the beach, too. On many accessible beaches worldwide, people like to picnic on the sand in front of their cars, making use of the headlights to illuminate the party in the evening, the car stereo to make music, and the battery to power their cellphones. And just like beach furniture is a nucleus of garbage production, cars are litter hotspots: rarely are only tire tracks left behind. Finally, illegal dumps near the sea always have their share of old car parts, batteries, tires, and the like. Much of it is one storm away from being washed into sea. If you find this all to be exaggerated and hard to believe, then read on.

Fig. 3.50 Vehicles on beaches? You'd be surprised what goes on in the preseason when beaches are cleaned and "landscaped" for your convenience. The little boy on the right is learning by doing. Mediterranean, Italy

Fig. 3.51 There are more than one billion vehicles on our planet; no wonder you find them being driven, parked, stuck, or abandoned on beaches. Spending the night on the beach rather than at a truck stop got this driver stuck. Mediterranean, Turkey

Fig. 3.52 These cars were brought to the beach as decorative elements for a beach disco. Mediterranean, Turkey

Fig. 3.53 This car has been given the fine retirement sea view it deserves. Caribbean, Guadeloupe

Fig. 3.54 Abandoned vehicles get ugly fast in the salty beach environment. Pacific, Mexico

Fig. 3.55 and Fig. 3.56 Two rusting vehicle frames dumped on the upper beach. Pacific, Mexico.
Either a truly horrific accident or total disregard for the beach environment.

Fig. 3.57 Car engines are formidable, long-lived beach debris. Caribbean, Grenada

Fig. 3.58 This engine is being used as a "cleat" to secure a boat. Caribbean, Grenada

Fig. 3.59 and Fig. 3.60 Where you find motor vehicles, you'll also find various parts such as batteries (treat as hazardous waste during beach cleanups!). Mediterranean, Ibiza.
If you find battery parts like this vent cap, the battery itself may be lurking just around the corner. Mediterranean, Turkey

Fig. 3.61 and Fig. 3.62 Car fenders are rarely metal these days, but sturdy marine debris nonetheless. Mediterranean, Turkey.
The same holds true for hubcaps. Mediterranean, Turkey

Fig. 3.63 Every vehicle has an oil filter which you can exchange (and discard) yourself! The metal housing rusts away to reveal the long-lived, oil-sodden filter element. Mediterranean, Turkey

Fig. 3.64 Complete your license plate collection on the beach. Check out the Internet for many fun recycling and upcycling ideas. Mediterranean, Turkey

Fig. 3.65 All-terrain freaks tend to lose parts when carving up beaches and dunes in testing the capabilities of their expensive gas guzzlers. Atlantic, England

Fig. 3.66 As beach litter, bicycles quickly lose their green alternative transportation aura. One recent international coastal cleanup yielded 435 bicycles. Pacific, Mexico

Fig. 3.67 Bicycle owners tend to decompose faster than their steeds though. Pacific, California

Fig. 3.68 Bicycle rims and spokes are built to take a beating. This means long-lived beach litter. Pacific, Mexico

Metal, Vehicles, and Tires 75

Fig. 3.69 This lock's bicycle-protection days are over. Mediterranean, Turkey

Fig. 3.70 This appears to be the track of a bulldozer. Massive beach litter but ultimately still no match for seawater. Caribbean, Grenada

Fig. 3.71 And believe it or not, sometimes you'll even find the abandoned track vehicle itself, although this one is admittedly on quite a remote beach. Antarctic, Deception Island

3.3 Tires

How many cars and trucks have been produced over the past 20 years alone? One rough estimate is one billion. That's four billion tires out of the showroom alone. Multiply that times let's say two sets of tires per vehicle lifetime, add winter tires for those of us who don't live in the tropics, and then tack on lots of motorcycles and scooters. That's an enormous number with many, many zeros. To paraphrase an old folk song, "where have all the tires gone?" Unfortunately, no one has yet come up with a truly gainful solution to the problem of used tires – at least not for the staggering amounts piling up around the globe. Importantly, tires are not simply "rubber" but contain numerous materials that can include nylon, fabric, Kevlar, and steel in highly intricate patterns. Melt tires down? Wrong material combination. Burn them – outrageous air pollution. Grind them down to powder – restricted demand (and the metals from "steel-belted" models require complicated pretreatment). What else can be done with tires? How about a little upscaling? Back in the day, there wasn't a hip shoe store that didn't stock sandals made out of old tire treads. Take the time and check out the Internet for dozens of amazing craft ideas and creative projects to recycle and upcycle discarded tires. The vast majority, however, end up in dumps and landfills.

Fig. 3.72 Where have all the tires gone? How about dumped by the thousands in this lagoon a stone's throw from the sea. Water-filled tires are mosquito breeding hotspots. Caribbean, Grenada

The world's largest tire dump is apparently in the Kuwaiti desert and contains seven million tires [8]. This tire graveyard is billed as being visible from space, although that designation has lost some of its bite considering that pretty much everything is visible from space these days.

You'll find tires on nearly every beach, too. The 25-year anniversary (2011) tally of the international coastal cleanups yielded 979,468 tires [9]. How do they get there? Some, strangely enough, are misguidedly positioned for recreational purposes (Fig. 3.73). Beyond the usual dumps too close to the shoreline or folks who discard them "on the cheap," two additional sources that may come as a surprise deserve mention here.

The first involves commercial boats: you'll often see them equipped with tires hung overboard on both sides (that would be port- and starboard). These so-called fenders serve as bumpers when docking at a pier or as protection when boats anchor side by side in harbors (Fig. 3.75). Some harbors line their walls with tires for double protection. They are ugly but cheap, practical, effective and a form of upcycling too! As you can imagine, such tires take a terrific beating, and it's usually the ropes or chains holding them in place that

Fig. 3.73 These tires – part of a children's playground – are only one storm away from becoming colorful marine debris. Mediterranean, Turkey

give away first. You can often recognize this source on the beach because either part of the rope is still attached or the holes bored into the tires for the tethering are visible.

The second unusual source involves "artificial reefs." This is a hot issue these days when natural reefs are being destroyed and man-made reef objects are built and sunk on the seafloor to provide shelter and habitat for marine life. A great excuse for the oil industry, for example, to save tons of cash and leave their only partially dismantled oil rigs and other facilities anchored to the seafloor as "habitat." These structures do attract fishes (lots of enthusiastic sport fishers!), but they have also been an all too common excuse for sinking a wide array of ugly items into the sea. This includes everything from decommissioned ships or old New York City subway train cars. What does this have to do with tires? The brilliant idea was to kill two birds with one stone: get rid of old tires and "help" the marine ecosystem with artificial reefs. On the East Coast of the USA, for example, millions of tires were sunk offshore, variously latched together and anchored to the bottom. It's not nice, however, to fool Mother Nature. And revenge she did take. The eco-designers clearly underestimated the power of the sea. Major storms and hurricanes easily uprooted and tore these structures apart. The result: loose tires rolling along the shallow seafloor with the tides and currents, crushing living reefs, sea grass beds, and all other creatures that got in the way. And wave action eventually washed many of these tires up onto beaches. The final tally is between one and two million tires sunk in Florida waters alone; several hundred thousand have already been retrieved (in part with military support [10]), and tens of millions of dollars will have to be spent by several states to collect these derelict ecological steamrollers [11].

What else do tires have working against them? Foremost, they are exceedingly tough. Built to take a horrendous beating for years on end, they can decide between life and death of humans travelling at speed. This means they are among the longest-lasting marine debris, second perhaps only to glass. And the additives used to keep them pliable are not necessarily environmentally friendly. That's one of the reasons you'll almost never see coral colonies or other organisms growing on tires. Beyond all the damage that tires do to the marine environment, they pose a human health hazard as well. Some tires are discarded along with their rims, which means they contain air and float very well. You don't want to hit one with your speedboat. Many tires have internal metal bands and cables that help stabilize their shape under tons of car or truck, but that form needle-sharp, rusty projections when exposed. On beaches, most are partially buried and filled with sand, so keep an eye out for these sharp metal components when digging them out. Wet-sand-filled tires are also many times heavier than you would expect: watch your back when prying them loose during beach cleanups.

Interested beach detectives can gain lots of information about tires from the sidewall markings. These provide the manufacturer's name, the tire model, production date and location, construction type, dimensions, load and speed ratings, pressure requirements, and much more. See the graphic below (Fig. 3.74) for a quick overview or enter "tire markings" or "tire codes" into your computer search engine for a complete rundown on how to read all the codes, symbols, and indexes!

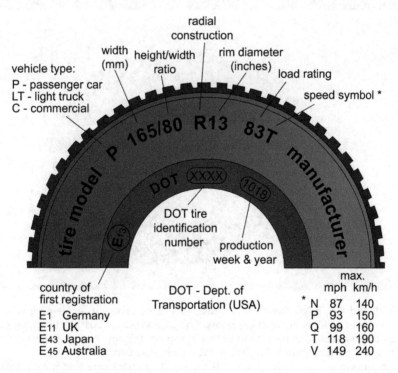

Fig. 3.74 Tire (sidewall) markings. More difficult to decipher than Egyptian hieroglyphics, but every tire tells a story!

Fig. 3.75 and Fig. 3.76 Old tires can take lots of punishment, explaining why they often serve as fenders on working boats. Rough seas and robust docking maneuvers can quickly make marine debris out of this upscaling strategy. Pleasure boats use more colorful, chic, and expensive fenders. Persian Gulf, Abu Dhabi.
A strap around or hole through a tire is the beachcomber's clue that it recently hung overboard as a cheap boat fender. Aqaba, Red Sea

Fig. 3.77 Ropes and chains are the weak link in any tire boat fender setup. Mediterranean, Turkey

Fig. 3.78 Imagine the forces that tore this tire fender – still attached to rusty chain remnants – apart. Atlantic, Florida

Fig. 3.79 Every beach detective eventually gets "tired." Air-filled tires explain how some large metal items (rims) can cross vast stretches of ocean, posing a threat to boaters, coral reefs, and other shallow marine ecosystems. Caribbean, Guadeloupe

Fig. 3.80 Sidewall markings tell beach detectives almost everything but the owner's name. Here, from left to right, width in millimeters (165), height-to-width (aspect) ratio (80), radial construction (R), rim diameter code (13), load index (83) and speed rating (T), and country where the tire was first registered (E mark in circle: E13 = Luxembourg). See Fig. 3.74 or find all symbols, codes, etc., on the Internet under "tire codes" or "tire markings".

Fig. 3.81 The heavier-duty the vehicle, the more massive the tire and the longer it remains as marine debris. Mediterranean, Turkey

Fig. 3.82 White tubeworms in the profile mean this tire spent time underwater. Tires quickly fill with sand or gravel, making beach cleanups back-breaking work! Mediterranean, Turkey

Fig. 3.83 Dense encrustation (barnacles) means time spent at sea as marine debris and also means an older tire whose toxic chemical additives have mostly leached out. Caribbean, Guadeloupe

Fig. 3.84 Metal rims add to the destructive force of loose tires on shorelines. Here, lodged in a coastal cliff along with tarred sheet metal and a gas cartridge. Marine debris hammered into rocky shores by waves can be difficult to dislodge. Mediterranean, Spain, Ibiza

Fig. 3.85 The calendars of some tire companies may be sexy, but their products as beach litter are certainly not. Mediterranean, Turkey

Fig. 3.86 and Fig. 3.87 Tires mean rubber… and metal, for example, high-strength steel cable in the so-called bead. Marine debris doesn't get much tougher than this. Watch out when digging tires out of the sand! Ibiza, Spain.
A few rubber remnants help beach detectives to reconstruct this metallic crime scene as being tire-related. Mediterranean, Turkey

Fig. 3.88 Not even bicyclists are always entirely environmentally friendly. Enter "upcycle bicycle tires" to find dozens of clever ideas for old tires and inner tubes on the Internet. Pacific, Mexico

Fig. 3.89 A stranded deep-sea monster? No, just a severely maltreated tire. Mediterranean, Turkey

Fig. 3.90 Number 56 out of 101 things you shouldn't do with old tires on beaches. Caribbean, Grenada

References

1. Frid CLJ, Caswell BA (2017) Marine Pollution. Oxford University Press, Oxford 268 pp
2. TIME International (1993) Landmines: the devil's seed. No. 50, 13 December 1993
3. Ottawa Treaty. https://en.wikipedia.org/wiki/Ottawa_Treaty
4. Harley-Davidson Museum. https://www.harley-davidson.com/us/en/museum/explore/exhibits/tsunami-bike.html
5. Weis JS (2015) Marine Pollution: what everyone needs to know. Oxford University Press, Oxford 273 pp
6. Metal upcycle recycle. https://pinterest.at/ariaism/metal-upcycle-recycle/
7. Marine debris is everyone's problem (2015) https://web.whoi.edu/seagrant/wp-content/uploads/sites/24/2015/04/Marine-Debris-Poster_FINAL.pdf
8. World's biggest tire graveyard in Sulabiya, Kuwait (2015) http://www.amusingplanet.com/2015/01/worlds-biggest-tire-graveyard-in.html
9. Ocean Conservancy. https://oceanconservancy.org/wp-content/uploads/2017/04/2011-Ocean-Conservancy-ICC-Report.pdf
10. https://en.wikipedia.org/wiki/Osborne_Reef
11. The Guardian (2015) https://www.theguardian.com/us-news/2015/may/22/florida-retrieving-700000-tires-after-failed-bid-to-create-artifical-reef

4

Plastic

In the movie classic "The Graduate," what was the one word that Benjamin Braddock (alias Dustin Hoffman) was told by a well-meaning family friend would ensure his career and future happiness? Correct, that word was "plastic." A mere few decades ago, it was hailed as the solution to virtually every problem facing humankind. From today's perspective, can we say anything nice about plastic? Well, for one it did, thankfully, replace certain natural products that required killing whales (corset supports, specialty oils), elephants (ivory piano keys), and sea turtles (tortoiseshell combs). And some folks might argue that it is a "natural product" – made mostly of crude oil. After all, making plastic products from oil – products that could potentially serve us for years or decades – is a better use of this finite fossil resource than simply igniting it in combustion engines to power our cars.

That said, being "originally natural" does not necessarily mean healthy for the environment. Just like spilled (natural) crude oil poses a major threat to marine ecosystems, improperly discarded plastic products have become a marine ecology disaster. From being hailed as a solution to every problem, plastic has now itself become one of the major problems facing the environment [1]. Nowhere is this more evident than in the marine debris issue – in fact, plastic and marine debris have become synonymous for most folks.

The weights and volumes of plastic produced every year are so staggering that the human mind has difficulty fully comprehending them. Multiply that by half a century of mass production. And all of it eventually becomes waste:

somewhere around 300 million metric tons of plastic waste are produced every year, with an estimated 8 million tons entering the oceans annually [2]. You'll find virtually every type of plastic item ever produced floating on the world's oceans, littering the seafloor, and washed ashore on beaches. And in the case of microplastics (see below), the smallest fragments have entered the food chain and made their way into our own bodies. This supports the contention of experts that almost every type of pollution on the planet ultimately becomes a marine pollution issue. There are two basic sources. This first is the items we use in our daily lives as landlubbers and that end up in the sea due to improper disposal, "accidents," and ignorance. The second is items placed in the ocean on purpose: the fact that plastic can be made to resist the harshest conditions, doesn't rust, and isn't usually biodegradable means that many products specifically designed for use at sea are made of plastic. This includes everything from fishing lines and nets to buoys and pleasure craft hulls. Simply put, plastic items are by far the most common marine debris and beach litter category you will encounter and make up the bulk of all the items collected during beach cleanups [3].

Plastic is not plastic. All types are made of long chains of giant molecules known as polymers (which is why the names of most begin with "poly-"). The final products themselves, however, span the full range from extremely rigid to very pliable and from feathery light to massive and impact-resistant. This reflects a multitude of basic components, many clever high-tech production processes, and untold chemical additives that impart additional features and colors [4]. Altogether this yields somewhere around 200 different plastics with often difficult-to-pronounce names such as polyethylene terephthalate, polymethyl methacrylate, or polyphenylene terephthalamide. To the initiated, these names more or less spell out what the respective plastics are made of. But because they are a mouthful for most of us, many are also known by abbreviations or acronyms. PE, for example, stands for polyethylene and PVC for polyvinyl chloride. Trade names are also commonly used, for example, Nylon® for polyamides or Kevlar® for polyphenylene terephthalamide. The important thing for beachcombers at the crime scene beach is that these abbreviations are usually stamped on products and enclosed in triangular symbols or "recycling" codes. Eighty percent of all plastics are so-called standard plastics that comprise most of the products we use in daily life. Such standard plastics include PE, PP, PS, PVC, and PET (see Fig. 4.1). They are used to make everything from bottles, shopping bags, pipes, automobile bumpers, carpet yarn, yoghurt cups, insulation panels, window frames, or synthetic fibers for clothing.

Fig. 4.1 The most common codes you'll find written on, stamped into, or embossed on plastic marine debris items

Why is plastic such a problem as marine debris? Simple: It exemplifies the whole issue because it doesn't go away, threatens sea life, poses serious hazards to humans at sea, is an esthetic nightmare, and reveals essential and unpleasant facts about human attitudes, foibles, and behavior. Accordingly, Nature Conservancy's top 10 list of dangerous debris items collected worldwide are all "plastic": bags, balloons, crab/lobster/fish traps, fishing line, fishing nets, plastic sheeting, rope, six-pack holders, strapping bands, and syringes [5].

Let's start with marine organisms. Plastics threaten biodiversity and marine ecosystems because (1) they are ingested and clog digestive tracts; (2) they entangle, maim, and drown organisms; (3) the chemicals they contain are marine pollutants; and (4) they can transport "alien" species to faraway places [6].

Ingestion? Sea life (and most life, come to think of it) is designed to inspect and think about consuming almost everything it encounters. Although plastic is clearly not on any evolutionarily honed food-detection program, it is consumed nonetheless. This is because filter-feeding organisms never evolved to distinguish between the plankton they normally consume and the equally sized, tiniest marine debris items (microplastics, see below) now so abundantly mixed into the "marine soup." What marine life filter-feeds? Most of it. This goes beyond the sponges, worms, mussels, sea squirts, and many other invertebrates to include the biggest filter-feeding organisms on our planet, namely baleen whales. Recent studies on fin whales in the Mediterranean and the Pacific, for example, have shown that they apparently filter out enormous amounts of microplastics along with their plankton diet: their tissues contained measurable amounts of plastics-related compounds, including additives.

Predatory animals are also duped by plastic. This is most obvious in the case in fishing lures and the like, which are specifically designed to trick our quarry into biting (see Chap. 11). In the case of debris, they may simply mistake plastic items for legitimate prey. This is probably the case with sea turtles and transparent plastic bags: any scuba diver will tell you that you have to look twice to distinguish a suspiciously quavering bag from a pulsating jellyfish. On the other hand, the turtles may simply be biting at the organisms that encrust many floating plastic objects, unwittingly swallowing the plastic as well [7]. One recent scientific paper determined that certain chemicals in plastics smell good to seabirds [8]. The result is the same in any case: plastic fills or otherwise blocks the digestive tract, leading to slow death by starvation. More than 90% of all seabirds, for example, have eaten plastic, and that number is expected to reach 99% by the year 2050 [9]. Decomposed bird cadavers on shores sometimes leave behind a small pile of plastic items inside the rib cage where the stomach once was (enter "bird cadaver plastic" into your computer's search engine). As with the largest filter-feeders, the largest predators on Earth, sperm whales, have also been killed by swallowing large plastic items, including the giant plastic sheeting associated with greenhouse agriculture [10].

The second sea life-related threat is entanglement. The range of hazardous items includes abandoned or lost fishing lines and nets, packaging material along with strapping bands, bags and six-pack rings. The most common victims are seabirds, sea turtles, seals, and dolphins. Such marine debris reduces its victim's mobility and causes starvation, strangulation and mutilation (lines cutting through and amputating sea turtle flippers), infection, and slow death. Even "slippery" fish can become entangled (in fact, that's the design principle behind "gill" nets). Larger fishery-related debris has become an entanglement threat even to the largest animals on the planet, namely whales. The International Whaling Commission, for example, has begun devoting considerable efforts to tackling this problem by establishing an entanglement response network [11]. The third threat is pollution by the many compounds and additives that slowly leach out of plastic as it ages. The fourth is the "rafting" of encrusting and other clinging organisms to faraway ecosystems on floating plastic items. This is one of the several mechanisms behind the increasingly alarming "invasive alien species" problem, in which artificially introduced species wreak havoc on an unsuspecting native fauna and flora. And plastic is ideal because it is so abundant, is not slippery like glass, and doesn't rot like wood.

Finally, for those who might be less interested in animal welfare, plastic marine debris poses a considerable hazard to humans and their activities at sea. Why should all divers and snorkelers have a very sharp knife readily at hand? Not to maltreat the rare inquisitive shark, but to be able to free yourself from unseen lines, nets, and other plastic debris underwater before you run out of breath or your air tank becomes empty. From the economic perspective, the world's pleasure boats and commercial fleets lose enormous sums of money every year from plastic that entangles propellers and propeller shafts, blocks cooling water intakes (quickly causing engine meltdowns), and clogs fishing nets. From a human health perspective, plastics slowly leach out their chemical additives and adsorb other pollutants at the water surface. The tiniest microplastics make their way into the tissues of the seafood we eat and ultimately into our own bodies.

Microplastics
The new catchword among marine debris specialists these days is "microplastics" [12], and everyone from local authorities to the United Nations has gotten into the act [13]. As the name indicates, this category comprises smaller plastic items, usually understood as being under 5 mm (for you remaining hold-outs of the Imperial and US customary measurement system, that's 0.19685 or 13/64 inches). Other experts consider the microplastic size class to begin at 1 mm, with sizes going down to the micrometer scale. (Just for the sake of completeness, researchers also distinguish even smaller particles in the nanometer range, nanoplastics, but this is not a marine debris field guide issue.)

Microplastics come in two major categories:

1. Primary microplastics: plastics specifically manufactured to be small such as plastic pellets, down to the microscopic "scrubber" particles in cleansers and cosmetics. Plastic resin pellets or "nurdles" are the initial product and starting material for the entire plastic industry (Figs. 4.2 and 4.3). They are then poured into forms to make the desired product. These pellets are themselves a major source of plastic pollution in the seas – one that is just

Fig. 4.2 The mother of all plastics: the resin pellets or "nurdles" poured into the forms used to make most plastic items. Mediterranean, Spain, Ibiza

Fig. 4.3 Pellets are instantaneous, so-called primary microplastics because they are produced to be small. Look closely along the high-tide mark. Make it a competition: see who can collect the most! Mediterranean, Spain, Ibiza

large and abundant enough for beach detectives to quickly recognize. Moreover, they contain enough toxic ingredients (POPs (persistent organic pollutants), which include DDT and PCBs) that they have been singled out for an "International Pellet Watch" [14]. Pellets on seashores mean that plastic is being massively lost into the environment even before the plastic products themselves are created. The problem starts far upstream in factories and plants along rivers. One recent study defined numerous hotspots along a nearly thousand kilometer stretch of the Rhine River as it meandered through several countries on its way to the North Sea [15]. Another found more microplastics in the Danube River near Vienna than fish larvae [16]. A shameful situation and clear evidence for sloppy practices and careless disregard for the environment by industry.
2. Secondary microplastics: the breakdown products of all larger plastic items. This includes everything from tiny remnants of plastic cups and bags to the thousands of tiny fibers released every time laundry containing clothes with synthetic fibers is washed. The estimates for every single washing of one piece of such clothing range from 10,000 to 250,000 fibers [17]. Importantly, even these small particles continue to leach toxic substances into the water. And they also absorb other pollutants from the ocean surface and funnel them into the food chain in concentrated form. These tiniest particles are then consumed and stored in animal (and in our own) tissues. They have already been detected in the oh-so-all-natural sea salt produced from the evaporation of seawater.

The abundance of microplastics, coupled with overfishing, means that, already today, there are probably more plastic items in the ocean than fish. In fact, the term "plastisphere" (unpleasantly alluding to "biosphere") has been coined to describe the microbial communities inhabiting plastic marine debris [18]. This means that, invisible to the naked eye, a thin film of life covers every piece of plastic (minimum estimate more than 5 trillion particles [19]) in the sea. Some scientists hope that such bacteria will speed up the plastic decomposition process. In fact, the oceans contain many times the above "trillions" number, but not necessarily floating on the surface. Discomfortingly, the actual fate of that invisible part is still being debated [20].

So, are we against plastic? If yes, what are the alternatives without going back to whalebone, ivory, and tortoiseshell? One idea is to make plastics out of raw materials other than oil, gas, or coal. It can actually also be made from other organic materials such as wood, corn, potatoes, and even algae. These can be grown on farms and in forests and are therefore renewable. Some of these products are more biodegradable than plastics made from crude oil. Finally, new plastic products can be made from plastic wastes such as cleaned plastic bottles, yoghurt cups, and the like. This is the recycling variant. More generally, the 6 "R"s (and the "U") apply here: *rethink, refuse, reduce, reuse, repair*, and *recycle*. Do I really need this product, and does it have to be made of plastic (rethink)? Do I really need to have every item I buy put into a separate plastic bag (refuse)? Do I have to immediately throw away the new plastic bag when I return home (reuse)? Do I really have to trash a product because a small plastic component is damaged (repair)? When I do finally discard a plastic item, is there a special bin I can put it (recycle)? Maybe I can use the old product to make a completely new item (*upcycle*).

Unfortunately, recycling is where theory and practice remain to be fully reconciled. First of all, voluminous plastic waste complicates storage and transport. Then, the many types of plastic stymie recycling programs because mixing them up does not necessarily yield the ingredients that industry can use. The result: plastic waste is often simply burned or, to make it sound more tech-savvy and sophisticated, "thermally recycled." Never do this yourself after a cleanup or at a beach campfire: beyond producing toxic gases, plastic melts together with sand, wood, and other debris to form a newly described kind of stone – "plastiglomerate" [21]. This material may form the future marker horizon for human pollution and the geological onset of the Anthropocene as the new era in Earth history. Only geologists love it.

The strategies to deal with plastic wastes vary from country to country, from state to state, and from city to city. They range from banning plastic bags to shipping plastic abroad on a large scale (e.g., China's very recent decision to ban the import of plastic wastes caused quite a gulp in the EU [22]). This is where the symbols stamped into or onto plastic products can play an important role. Such "recycling codes" (Fig. 4.1) will one day hopefully more strongly promote targeted collection and recycling efforts, reducing the amount that is burned or dumped into landfills and that enters the sea.

For the sake of simplification in this guide, the enormous scope of plastic items is treated under separate, more intuitive subheadings such as cups, bags, toys, and packaging or in separate chapters such as styrofoam, personal hygiene (combs to shavers), medical (syringes), fishing (nets to buoys), or cigarettes (yes, the filters are largely plastic!). This is also where you'll find more specific reduction or recycling tips. Of course, most plastic items found on beaches also contain other components such as metal or glass and may be included under those headings if those other parts dominate.

References

1. Derraik JGB (2002) The pollution of the marine environment by plastic debris: a review. Mar Pollut Bull 44:842–852
2. Ocean Conservancy. Fighting for trash free seas. https://oceanconservancy.org/trash-free-seas/plastics-in-the-ocean/
3. Ocean Conservancy https://oceanconservancy.org/trash-free-seas/international-coastal-cleanup/annual-data-release/
4. Plastic Additive Standards Guide. AccuStandard (2015) https://www.accustandard.com/assets/Plastic_Add_Guide.pdf
5. Ocean Conservancy act.oceanconservancy.org/site/DocServer/MarineDebris.pdf?docID=4441
6. Secretariat of the Convention on Biological Diversity and the Scientific and Technical Advisory Panel– GEF (2012) Impacts of marine debris on biodiversity: current status and potential solutions. Montreal, Technical Series No. 67. 61 pp
7. Nelms SE, Duncan EN et al (2015) Plastic and marine turtles: a review and call for research. ICES J Mar Sci. https://doi.org/10.1093/icesjms/fsv165
8. Savoca MS, Wohlfeil ME et al (2016) Marine plastic debris emits a keystone infochemical for olfactory foraging seabirds. Sci Adv 2(11):e1600395. https://doi.org/10.1126/sciadv.1600395
9. Wilcox C, Van Sebille E, Hardesty BD (2016) Threat of plastic pollution to seabirds is global, pervasive, and increasing. PNAS 112(38):11899–11904 https://doi.org/10.1073/pnas.1502108112
10. de Stephanis R, Gimenez J et al (2013) As main meal for sperm whales: plastic debris. Mar Pollut Bull 69:206–214 https://doi.org/10.1016/j.marpolbul.2013.01.033
11. International Whaling Commission https://iwc.int/entanglement-response-network
12. Cole M, Lindeque P et al (2011) Microplastics as contaminants in the marine environment: a review. Mar Pollut Bull 62(12):2588–2597 https://doi.org/10.1016/j.marpolbul.2011.09.025
13. UNEP (2015) Resolution UN/EA-1/6 Marine Plastic Debris and Microplastics. http://www.un.org/depts/los/consultative_process/ICP17_Presentations/Savelli.pdf
14. Heskett M, Takada H et al (2012) Measurement of Persistent Organic Pollutants (POPs) in plastic resin pellets from remote islands: toward establishment of background concentrations for International Pellet Watch. Mar Pollut Bull 64(2):445–448 https://doi.org/10.1016/j.marpolbul.2011.11.004
15. Mani T, Hauk A et al (2015) Microplastics profile along the Rhine River. Sci Rep 5:17988. https://doi.org/10.1038/srep17988
16. Lechner A, Ramler D (2015) The discharge of certain amounts of industrial microplastic from a production plant into the River Danube is permitted by the

Austrian legislation. Environ Pollut 200:159–160 https://doi.org/10.1016/j.envpol.2015.02.019
17. Pirc U, Vidmar M et al (2016) Emissions of microplastic fibers from microfiber fleece during domestic washing. Environ Sci Pollut Res Int 23(21):22206–22211. https://doi.org/10.1007/s11356-016-7703-0
18. Zettler ER, Mincer TJ, Amaral-Zettler LA (2013) Life in the "Plastisphere": microbial communities on plastic marine debris. Environ Sci Technol 47:7137–7146. https://doi.org/10.1021/es401288x
19. Eriksen M, Lebreton LCM et al (2014) Plastic pollution in the World's oceans: more than 5 trillion plastic pieces weighing over 250,000 tons afloat at sea. PLoS One 9(12):e111913 https://doi.org/10.1371/journal.pone.0111913
20. Thompson RC, Olsen Y et al (2004) Lost at sea: where is all the plastic? Science 304:838
21. Corcoran PL, Moore CJ, Jazvac K (2014) An anthropogenic marker horizon in the future rock record. GSA Today 26(6):4–8. https://doi.org/10.1130/GSAT-G198A.1
22. EU targets recycling as China bans plastic waste imports https://www.reuters.com/article/us-eu-environment/eu-targets-recycling-as-china-bans-plastic-waste-imports-idUSKBN1F51SP

4.1   Plastic Beverage Containers and Co.

Plastic beverage containers are among the most common and visible types of marine debris. Yes, we are supposed to drink plenty of fluids on the beach, and, yes, this means bringing plenty of drinks in the cooler. But why aren't these containers taken back home? Maybe they have become soda-pop sticky, sunscreen oily, and stuck-in-the-beach sandy or were used as an ashtray and have cigarette butts swirling in disgusting, warm soda drink remnants. There is always an excuse. And throwing empty containers over your shoulder has apparently not been declared a no-no in some cultures or has at least not filtered down into the consciousness of certain generations. And, if the beverages are alcoholic, intoxication is usually counterproductive for any activity, including conscientious recycling efforts.

You'll find beverage containers and related paraphernalia – bottles, drinking cartons, cups, six-pack holders, and straws – everywhere: rolling in the surf, lying on the beach, stuck into the sand, and blown back into the dunes. And you'll almost never find them alone. This is because most people need more than one drink during a day spent at the beach. On the one hand, ever smaller and cutesier drinks mean more packaging for ever shrinking, "personalized" portions. On the other hand, supersized drinks mean sturdier, more long-lived containers and more cups to provision the whole family or party.

Multiply this times a beach packed with visitors, add wind, factor in the conduct of sunburned and tired people beating a hasty retreat at sunset, and then consider the mostly under-dimensioned, suboptimally designed and intermittently emptied garbage bins at beach exits. It's a wonder you don't find even more beverage containers than you do. The bottles themselves are only the most visible marine debris in this category. Every bottle has a label and a cap, every cap a retainer ring. Many cups have a lid, and most lids come with a straw or plastic stirrer. Finally, many drinks are purchased with and held together by plastic "six-pack rings."

Plastic Bottles Wonderfully sculptured, with sleek necks, rounded shoulders, and sculpted lips. No, we are not talking about beautiful women but about the features of most plastic bottles. They have increasingly replaced the traditional glass bottle because they are lighter weight and unbreakable. The number of beverages they can contain is countless. The worst offender, however, is water: clean drinking water is a scarce commodity in many regions of the world, and water in plastic bottles has become a key trustworthy source (even if it merely contains hyped tap water). Apparently, about 100 million such bottles are used worldwide *every day*. Every one is "single-use" and only 1 in 5 is recycled. This easily puts them way up on the top 10 list of items collected during beach cleanups every year: in a recent, single-day event, over 1.5 million were recorded [23] (2nd place after the consistent winner: cigarette butts!). And they do not degrade quickly, especially if buried in the sand (one estimate is a life span of 450 years [24]). What can you do? Refuse and replace them with recyclable or, even better, returnable and reusable glass bottles. When going to the beach (or elsewhere), fill your drink into a personal drinking bottle and leave the plastic container at home. Discard all plastic bottles in recycling bins, crushing them flat first and screwing on the cap. This keeps air out, reduces bulk for transportation, and ensures that the caps themselves don't become separate marine debris items. And you can't believe the many ways to upscale plastic bottles (enter the last three words into your computer's search engine)!

Labels Almost every bottle is embossed with information or has a label, often made of durable, tightly shrink-wrapped plastic. Such labels contain detailed information not only about the bottle's contents but on its material composition. For our purposes, they also contain recycling information. Unfortunately, most people don't read such "small print" or tend to ignore the accompanying symbols. Like on most products, you'll also be sure to find the proud beverage company's address. And even better, their email address and, today, Facebook and twitter contacts. Do you think this company has done all it can to prevent beach litter and marine debris? Has the beachside eatery made a great enough effort to make sure its cups, lids, straws, etc. are properly disposed of? If not, why don't you simply contact them and tell them so. You might even try to trigger a shitstorm! Never underestimate the power of a single person to make a difference.

Caps Caps, threaded and ribbed, are the thickest and most robust plastic parts of plastic beverage containers. Made to be squeezed tightly, broken free from their retainer rings or other "tamper-resistant" fittings, and repeatedly twisted in both directions, they clearly survive the longest. When the bottle is empty, the best solution – if it's not a deposit bottle – is to squeeze it flat, screw on the cap, and discard it in the appropriate plastic waste bin. Of course, the Internet is full of websites with ideas on arts-and-craftsy things to do with bottle caps, especially if you want to keep your schoolchildren busy for hours in a non-digital pastime. In some places, plastic bottle caps are collected in enormous numbers for charity purposes. The caps are recycled, and part of the proceeds goes to purchasing wheelchairs and other noble causes, although some such schemes have turned out to be hoaxes (enter "plastic bottle caps charity" into your computer's search engine).

Cups If you are thirsty and don't like to drink out of a bottle used by other people whose medical status you may or may not know well, you'll need cups. And chances are you'll need a few cups for every person because, honestly, who keeps track of and reuses his/her cup throughout a day at the beach. Each cup is destined to become a marine debris item unless you take dedicated action. And when you buy your takeout drink from the beach bar or restaurant, the fast-food chain, or the coffee store, chances are it will come with at least some of the following: a plasticized cup, a (plastic) lid, a (plastic) straw, a (plastic) stirrer, a (plastic) coffee creamer portion cup, or a (plastic) sachet of whitish granules or of manufactured white fluid to make your coffee whiter. As a beach sleuth, you can actually read most lids to find out what people are drinking these days, even without the cup itself. How? Many of the plastic lids have a ring of raised bubbles around the edges, each with an abbreviation for the drink inside. For soda drinks, these might include "COLA," "DIET," "RB" (root beer), "TEA," or "OTHERS." Since the drinks are indistinguishable from the outside, the person making the drink simply presses down the appropriate bubble so that the server, in turn, will know which cup contains which drink. In the case of hot beverages, the buttons might, for example, be designated "BLACK," "LATTE," "CAPP," or "CHOC," although the more exalted coffee drinker who wants a decaffeinated double cappuccino with dash of organic hazelnut cream and hot, nondairy, low-fat milk will be hard-pressed to find the respective button.

Six-Pack Holders The plastic rings designed to hold six beverage cans have long been a notorious marine debris item. This is because many of these holders were found strangling seabirds and disfiguring fishes and other marine and freshwater life. They are in the Ocean Conservancy's top 10 list of dangerous debris items collected worldwide. The 25-year total tally (back in 2011 [25]) was already 957,975! To counter this entanglement threat, designs have been introduced that reduce the possibility of an animal sticking its head through one of the holes. Other strategies have included making thinner, predetermined breaking points or producing the entire item out of degradable plastic. One brewery in Florida makes an edible six-pack ring.

Straws Straws are an integral part of every takeout drink "package," and they regularly make it into the top 10 list of items recovered during international beach cleanups (Nr. 7 in 2017) [23]. Although made of paper in the early days, most are plastic today. They come in all lengths, sizes, and colors and can be straight, have a crinkled bend, be variously twirled, and bear tinsel crowns or funny party figures. Most come wrapped individually in a separate plastic sheath, "doubling" the litter problem. Straws are a serious marine debris issue. According to one estimate, somewhere around 500 million of them are being used and discarded every day in the USA alone. More than 400,000 were collected during one of the last international beach cleanups. One recent Internet video showed the painful process of removing a straw from a sea turtle's nostril! Some cities have taken steps to prohibit beachside restaurants from using plastic straws. You can help by saying "no" to single-use, disposable straws. Or take it one step further and start a conversation with your local eatery (straws served only upon request/change to nonplastic options, etc.). What are your alternatives? Bring your own: you can actually buy reusable straws made out of silicone, stainless steel, or glass. Even organic versions made of bamboo or real "straw" are available. Check out the Internet for the many companies offering such items or for the many ideas (enter "upcycle/recycle straws") on how to create something artistic out of used straws!

References

23. Ocean Conservancy https://oceanconservancy.org/wp-content/uploads/2017/06/International-Coastal-Cleanup_2017-Report.pdf
24. Marine debris awareness poster https://web.whoi.edu/seagrant/outreach-education/marine-debris
25. Ocean Conservancy https://oceanconservancy.org/wp-content/uploads/2017/04/2011-Ocean-Conservancy-ICC-Report.pdf

Fig. 4.4 Plastic cups – discarded full, empty, or used as an ashtray – have no place on the beach

Fig. 4.5 and Fig. 4.6 Typical "frayed" deterioration stage. The reinforced drinking rims tend to survive the longest.
Sometimes you'll have to look twice to recognize a cup. Mediterranean, Sardinia

Fig. 4.7 The "law of marine debris aggregation" and human behavior mean you'll rarely find a lone cup. These were "weighted down" by diligent beachgoers. Mediterranean, Turkey

Fig. 4.8 Stacking cups, whether new or used, definitely delays their decomposition. Mediterranean, Turkey

Fig. 4.9 Lids and straws, the most typical drink-related items on beaches (No. 6 and 7 in the top 10 in a recent international beach cleanup). A pushed-in "button" can reveal the original drink. Pacific, California

Fig. 4.10 No matter how festive a party straw might be, it's still a top 10 marine debris item. Mediterranean, Turkey

Fig. 4.11 Some folks take their drinks very seriously: supersized cups mean supersized straws! Pacific, USA

Fig. 4.12 Supersized straws and their caps mean extra-strength, durable marine debris items (caliper opened to 1 cm). Pacific, USA

Fig. 4.13 Straws often come individually wrapped. This doubles the number of straw-related marine debris items and further slows degradation. Note tar (bottom, right) and bottle cap, showing typical litter densities (when you look close enough). Mediterranean, Greece

Fig. 4.14 They come in all sizes, shapes, and colors – and often get together for a party. Mediterranean, Turkey

Fig. 4.15 To borrow a phrase from the film "Zorba the Greek": bottle, cap, retainer ring, wrap label, and contents – the "full (beverage container) catastrophe." Mediterranean, Turkey

Fig. 4.16 Buried bottles can survive the longest, 450 years according to one estimate. Atlantic, USA

Fig. 4.17 With the rising morning sun, empty bottles glow like light bulbs. They can be spotted and counted from afar. Mediterranean, Turkey

Fig. 4.18 The thicker, structured bottom parts of plastic bottles tend to survive the longest …. Mediterranean, Turkey

Fig. 4.19 … especially when reinforced with an extra plastic "boot," which tends to retain its shape even when the rest of the bottle has become "compromised," as the materials experts like to say. Mediterranean, Turkey

Fig. 4.20 Heavy barnacle growth means lengthier time spent at sea as marine debris – and the threat of introducing alien species to faraway, unprepared habitats. Pacific, South Korea

Fig. 4.21 Although "reuse" is one of the laudable 6 "R"s, beverage bottles are sometimes misused for new, potentially hazardous contents, in this case probably oil. This bottle was secured somewhere with fishing line. Caution during beach cleanups – never open any bottle or canister! Mediterranean, Turkey

Fig. 4.22 Plastic caps are the thickest and most durable parts of any bottle, here still enclosing remnants of a glass bottle lip. Caution: even "harmless" items can cause injury. Mediterranean, Turkey

Fig. 4.23 The "law of marine debris aggregation" holds especially true for bottle caps, although some accumulations are more difficult to explain. Mediterranean, Turkey

Fig. 4.24 Robust plastic drinking bottle spouts have multiple movable parts and survive much longer than the bottles themselves. Mediterranean, Turkey

Fig. 4.25 Plastic inserts help barkeepers accurately dose a stiff cocktail but also make for very stiff marine debris. Mediterranean, Turkey

Fig. 4.26 Rarely does the "law of marine debris aggregation" apply more visibly than to plastic bottles. Once a few are left behind (or blown together by the wind), many folks feel free to add to the pile. Mediterranean, Turkey

Fig. 4.27 Tying used plastic bottles together for reuse as cheap buoys doesn't really help solve the marine debris problem. Caribbean, Cuba

Plastic 109

Fig. 4.28 Beverage cartons ("tetrapacks"): lightweight, space-saving, unbreakable, but made of paper, plastic and metal foil and therefore long-lived marine debris. Note detached plastic spout (top left). Mediterranean, Turkey

Fig. 4.29 and Fig. 4.30 Like most marine debris items: made by the hundreds of millions and recognizable even after considerable decay. Mediterranean, Turkey.
Sturdy spouts and caps survive the longest; here white paper and plastic "tetrapack" remnants still attached. Mediterranean, Turkey

Fig. 4.31 Labels are a plastic category of their own. Very durable and full of information about the plastic type and recyclability. For every label, there's a bottle (and a cap) lying around somewhere. Mediterranean, Turkey

Fig. 4.32 As a conscientious parent, you could help avoid the instant-drink-and-throw-away portions for kids. Pacific, USA

Plastic 111

Figs. 4.33, 4.34, and 4.35 Not to forget their many twist-off-and-drop-on-the-beach caps. Atlantic, USA

Fig. 4.36 Six-pack rings: a deadly trap for seabirds and other marine life. Collect and destroy them even if you are not doing a beach cleanup. Newer models may have pre-determined "weak links," be made of "biodegradable" plastic, or even be edible.

Fig. 4.37 and Fig. 4.38 Beach detectives sometimes have to look twice to recognize six-pack rings. The 25-year international beach cleanup total: nearly one million! Estimated decomposition time, 450 years. Pacific, USA.
The architectural features of even the smallest fragments can be a telltale clue. Pacific, USA

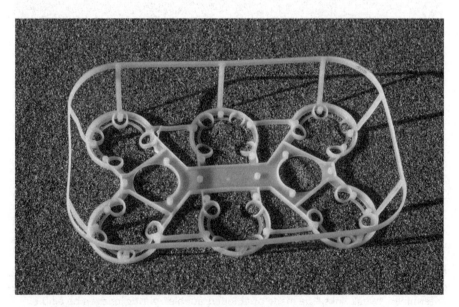

Fig. 4.39 The sturdier the design, the longer-lasting the marine debris. This holder is for bottles. Pacific, Mexico

Fig. 4.40 Six-pack holder design with fewer holes to ensnare sea life. Mediterranean, Turkey

Fig. 4.41 Crunching bottles can save lots of space. Why don't more companies design them to be easily scrunchable? Mediterranean, Turkey

4.2 Plastic Canisters

Plastic canisters are simply larger and more robust containers than most plastic beverage bottles. While some do contain drinks, many are filled with oil, chemicals, or cleaning agents. Regardless of their content, they are all built to withstand outside forces and safely carry a lot of weight. This means they are thicker-walled, sturdy, and resistant to degradation by their often caustic, toxic, or otherwise unpleasant contents. They are also usually equipped with various types of built-in handles or grips to simplify carrying the bulky product. Those tend to be particularly hardy litter. Product robustness is always a marine debris issue. On the one hand, such canisters decompose very slowly in the environment. On the other hand, their strength (and original expense) entices many folks to reuse them for other purposes. While any reuse does meet one of the 6 "R" criteria (rethink, refuse, reduce, reuse, repair, recycle), one of the most common reuses for boaters and fishers is as a cheap buoy. Even this apparently benign further life inevitably leads to a future marine debris item.

On the beach, you'll be able to recognize the original product in many canisters by their shapes and by their labels or, when these are eroded away, by embossed logos or addresses. During beach cleanups, caution is warranted: never open a canister yourself to check if it has contents and what those might be. First, no canister (or any other container for that matter) has ever been emptied literally to the last drop! Second, reused canisters mean that beachgoers may often be confronted with unexpected, potentially hazardous contents. Such canisters may be unlabeled or, even more disconcerting, bear their original labels but have entirely different contents. Don't blindly trust a label! If a large canister is partially buried in the sand or clearly isn't empty, call your beach cleanup supervisor to determine the right course of action: attempting to dislodge or move it yourself can cause aged, brittle, or damaged canisters to rupture and spill their contents on you and the beach.

Fig. 4.42 Canisters, many originally containing various motor oils, are the largest and among the potentially most hazardous plastic containers. And the "law of marine debris aggregation" often applies – you'll seldom find one alone. Atlantic, Scotland

Fig. 4.43 Large, barrel-sized canisters with unknown contents should be treated as hazardous wastes. Don't try to dislodge them on your own (and never open them). This one has been cut lengthwise (see also Fig. 4.57). Mediterranean, Turkey

Fig. 4.44 Most fresh canisters contain labels with lots of information…and at least some of their original contents. Remember – no liquid product (especially oils) was ever poured out "down to the last drop." Atlantic, Scotland

Fig. 4.45 If the labels are gone, you can often recognize canisters by embossed logos or shape. Mediterranean, Ibiza

Fig. 4.46 Many containers, like this wool detergent, can be recognized even without labels or embossed text. Mediterranean, Greece

Fig. 4.47 and Fig. 4.48 The embossing can tell beach detectives almost everything they need to know. Pacific, California.
… including the material ("2" means HDPE: high density polypropylene; see Fig. 4.1), and encrusting barnacles mean time spent at sea. Pacific, California

Fig. 4.49 When canisters degrade, the thinner walls are the first to become "compromised," as they like to say in businesses where things go wrong. Pacific, California

Fig. 4.50 The reinforced rims of plastic buckets, trays, and tubs make for tough marine debris. The walls can break or fray in patterns resembling their baby brethren, plastic cups. Mediterranean, Turkey

Fig. 4.51 Reinforced spouts, handles, and grips survive the longest. Mediterranean, Turkey

Figs. 4.52 Caribbean, Cuba

Fig. 4.53 and Fig. 4.54 For beach detectives, the paint and barnacle-covered rope around the handle point to reuse as a buoy. Caribbean, Cuba.
Rope and canisters – a common combination for cheap buoys around the world. Not every type of reuse helps avoid marine debris. Mediterranean, Turkey

Figs. 4.55 and 4.56 Many containers are camouflaged. Use caution when handling bleached and brittle canisters with potentially hazardous contents. Mediterranean, Italy

Fig. 4.57 Canisters carefully cut lengthwise are evidence for reuse for oil changes, a conclusion supported by oily residues as well. Caribbean, Cuba

Fig. 4.58 Various retainer or "tamper-proof" rings soon detach from container openings and caps. Large diameter (compare cigarette butt) points to a canister. Mediterranean, Turkey

Fig. 4.59 Only heavier containers such as canisters typically have extra ring grips. Mediterranean, Turkey

Fig. 4.60 ... which ultimately break off at joints, insertion points, or other weak links. Mediterranean, Turkey

4.3 Toys

What do children do on the beach all day? Right – play, play, and play some more. And what do they play with? Toys of course. The most common items are plastic sand toys. What can be more fun than digging a hole in the beach or building a sand castle? And this extends to "children of all ages," with young adults and parents often being the most ambitious and resolute sand diggers and sculptors. And for this, you need plastic shovels, rakes, cups, buckets, and sieves, along with many plastic molds to make sand cakes, sand figures, castle turrets, and the like. One or the other such item is bound to remain on the beach – accidentally buried, washed into the water by waves, caught by the wind, or simply left behind when beating a hasty retreat for lunch or as the sun sets in the evening. These toys typically come packaged in colorfully jumbled sets, meaning they aren't easily distinguishable from those belonging to other children playing nearby: better be on the safe side and leave them behind rather than be accused of stealing other children's toys!? Some flimsy plastic forms don't survive duty in the heavy wet sand, and who likes to take broken things back home? Finally, there is the ghastly fate of many floating toys that simply … float away … or that bob less dashingly than advertised and sink in the murky waves.

Many kids bring their favorite toys to the beach: a doll or a toy truck, for example. Why are these common debris items on beaches? They probably reflect a combination of exhausted, forgetful kids and clueless and overtaxed parents. By the time you realize something is missing, you are likely already at home many miles away, and it may well be dark. Going back at night or the next day is rarely an option: the chances of finding the exact spot where your family sat are poor. Moreover, what you didn't take home has no doubt been salvaged by someone else (why leave a perfectly intact toy lying on the beach if you're the last to leave or the first to come in the morning?). Ultimately, lost toys become eroded or disassembled in the surf, sandblasted by the wind, bleached and brittled by the sun, and crushed underfoot. Metal parts such as wheel axles of toy trucks rust away, leaving marooned, solitary plastic wheels. The long decomposition process to microplastics has begun.

In rare cases, new plastic toys have been "lost at sea" in bulk. A classic example is the famed toy duck marine debris event. (For the notorious 1990 accident involving sports shoes, see Chap. 9.) In 1992 a freighter bound from China to the USA lost part of its cargo, namely a container filled with 28,800 plastic bath toys, in the North Pacific. The toy ducks in the packages washed ashore all over the Pacific and, astoundingly, even reached the Atlantic (embedded in sea ice via the Arctic Ocean). Faded but still recognizable ducks were found in England after having traveled for nearly 20 years and covering more than 30,000 miles. Enter "toy duck Pacific" into your computer's search engine for detailed maps and many versions of the full story! Or read a book on the subject, "Moby Duck" [26].

Fig. 4.61 We all come to the beach to "play," in the broadest sense, and sand castles are at the top of the agenda. Mediterranean, Ibiza, Spain

Fig. 4.62 and Fig. 4.63 And building sand castles requires a full array of colorful tools. Mediterranean, Ibiza, Spain.
Toy tools and sand forms are easily lost, washed into the sea by waves, buried in the sand, or blown away by the wind. Mediterranean, Turkey

Fig. 4.64 Sieves are very useful both for play and to recover small or valuable items swallowed by the sand. Pacific, California

Fig. 4.65 Some sand forms bear important messages that any beach detective can interpret. Mediterranean, Turkey

Fig. 4.66 Like many plastic items, toys eventually become brittle and fragment, especially when crunched underfoot. Most folks won't bother with collecting all the pieces. Microplastics in the making! Mediterranean, Turkey

Fig. 4.67 Heavy wet sand or a misstep is more than many cheap toys can handle. Mediterranean, Turkey

Fig. 4.68 Plastic trucks are a classic beach toy – and common beach litter. Some kid somewhere is probably inconsolable. Mediterranean, Turkey

Fig. 4.69 Often only the more robust truck body itself remains. Wheels on beaches never survive long. Mediterranean, Turkey

Fig. 4.70 Wheels soon detach, and metal axles quickly rust, leaving behind many solitary plastic wheels as beach litter. Mediterranean, Turkey

Fig. 4.71 Larger wheeled children's toys don't fare much better on beaches, as this tricycle front fork attests to. It's a drama if you don't take them to the beach and can be a drama if you do. Mediterranean, Turkey

Fig. 4.72 Shoot 'em up – why should we expect the real world to make a halt at beaches. Mediterranean, Turkey

Fig. 4.73 What would a beach vacation be without a squirt gun. The more high-tech, the quicker they fall victim to the sand … Mediterranean, Turkey

Fig. 4.74 … exposing the internal parts and revealing that most complex plastic toys also have metal parts. Mediterranean, Turkey

Fig. 4.75 A small loss of body parts, a major step for marine debris. Mediterranean, Turkey

Fig. 4.76 The rough shore environment typically leaves disjointed or even somewhat risqué doll parts. Mediterranean, Turkey

Fig. 4.77 Many a tear may have flowed. Bringing your favorite doll to the beach can be risky. Mediterranean, Turkey

Fig. 4.78 What goes up must come down – but on windy beaches rarely where expected. This holds equally true for kites, balloons, fireworks, candle-powered "sky lanterns" and, more recently, drones. Red Sea, Jordan

Fig. 4.79 Plastic packaging such as this surprise toy capsule is a common marine debris item: there is often more plastic in the packaging than in the toy itself. Mediterranean, Turkey

Fig. 4.80 As plastic marine debris, many toys are as deadly to marine life as this one indicates. Mediterranean, Turkey

4.4 Balloons and Co.

Toy- or party balloons are a child's brightly colored – a simple, traditional, non-electronic source of joy. So it's all thumbs up, right? Well, they are also used in mass releases – usually printed with company logos or messages – at various publicity events. And what goes up must come down. Air-filled balloons can be blown about for a week or so before shriveling, although the helium in helium-filled balloons tends to escape faster. Many end up on the water and on beaches. Don't be lulled into believing that balloons, made of synthetic or natural rubber (latex), are biodegradable because they are somehow "natural": balloons may survive for months and longer on the water. During this time, floating balloons and their ribbon tethers can be swallowed by marine organisms (think sea turtles) and can entangle them (think seabirds).

Foil or mylar balloons – rubber balloons coated with an ultrathin metallized film – are a special case. These mirrorlike, often cutely shaped, and decorated gift items are easily recognizable (they don't stretch), retain helium much better, and do not fade or biodegrade. Their lightness and robustness mean long flights, and when they come down, they can interfere with power lines (power outages) or pose a threat to marine life. In one well-publicized case, a small whale needed several operations to remove swallowed marine debris: a mylar balloon was the largest object.

For decades, people have used balloons for fundraising and public relations events or to celebrate various festive occasions with mass balloon releases. These practices should be discouraged because they pose problems for the environment, wildlife, and cities. Releasing a balloon is merely "delayed littering"!

What can you do to help? Simply don't release balloons. If you want to jazz up your event, consider flags, reusable banners, streamers, tissue paper pom-poms, paper or fabric bunting, wind pinwheels/spinners, and the like. Blow bubbles instead of balloons. Send up retrievable kites. Support legislation that prompts consumers and retailers to tether helium-filled foil balloons with special weights. Enter "balloons and environment" into your computer search engine, and you'll be surprised at the number of websites and organizations devoted to educating the public about the hazards of releasing balloons. The simple motto – "don't let them go"!

References

26. Hohn D (2011) Moby Duck: the true story of 28,000 bath toys lost at sea and of the beachcombers, oceanographers, environmentalists, and fools, including the author, who went in search of them. Viking, New York 402 pp

Fig. 4.81 What goes up must come down. Knotted, gradually deflated, but still very durable marine debris. Pacific, Japan

Fig. 4.82 Never inflated, never knotted, pristine specimens are a bit more difficult to explain. Pacific, USA

Fig. 4.83 Killer jellyfish? No, balloon – how they break depends on the pressure inside. Pacific, Hawaii

Fig. 4.84 First degradation stage: mouthpiece still attached to part of balloon body. Atlantic, USA

Fig. 4.85 Final stage: thickest, mouthpiece part lasts the longest. Pacific, USA

Fig. 4.86 Like many marine debris items, balloons often come with "strings (or ribbons) attached"... Pacific, USA

Fig. 4.87 ... or with various plastic pieces to help prevent air or helium gas from escaping ...

Fig. 4.88 ... or even larger plastic components like this plastic inflator: orange rubber mouthpiece still visible. Atlantic, USA

Fig. 4.89 Glossy, reflective "mylar" balloons. Such novelty items are covered with a metalized film, like Christmas tinsel, and do not degrade or shred like traditional rubber balloons, making them more harmful to the environment. Mediterranean, Turkey

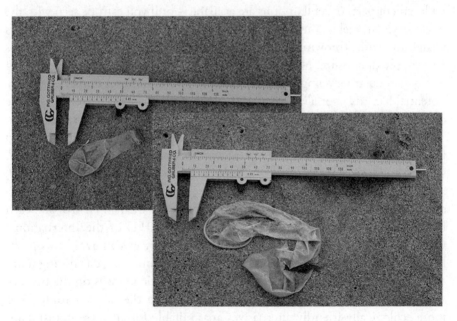

Figs. 4.90 and 4.91 A balloon and its imposter cousin, the condom: size does make a difference! (caliper open to 1 cm). Pacific, USA

4.5 Household Plastic

If you removed all the plastic items from your home, you'd probably be surprised at how empty it would be. Beyond the bottles, cups, bags, toys, sports equipment, and the many other categories treated here in separate chapters, almost every item we pick up, sit on, or listen to at home turns out to be made entirely or largely of plastic. Whether you are cooking in the kitchen, working in the office, combing your hair in the bathroom, or tinkering in the garage, chances are you are holding something plastic. This goes beyond the usual knickknacks to include everything from our patio furniture to the multimedia devices and the recording media we shove into them.

Household items are made to be used, and usage means wear and tear. It also means that many plastic items need to be replaced at regular intervals, often because some small part is defective and cannot be purchased separately. Planned obsolescence may also play a role, as does fashion or, more precisely, falling out of fashion. Many items are "single-use" in the sense that when the contents are used up, the whole product must be discarded: your ballpoint pen cartridge often cannot be replaced, the liquid in the spray bottle doesn't come as a refill extra, and the egg cartons aren't meant to be reused. If an item is broken or no replacement parts or refill options are available – or if you can't be seen with the out-of-style model – out the window it goes. Proper disposal isn't always straightforward. Throwing away whole items in the normal trash sometimes fails simply due to size. Many multicomponent products don't seem to belong in any particular recycling bin; others have remnant contents or parts that pose potential hazards. Not all unwanted household plastic items therefore take the correct waste disposal path, and a fair share ultimately lands in our oceans.

With regard to marine debris, perhaps the most direct route household plastic takes to the beach is in the picnic basket. Plastic plates, cups, forks, spoons, and knives are the staples of most serious outings. They are cheap and, after use, usually sticky, greasy, soggy, sandy, or otherwise icky – classical reasons for not taking them back home. Such tableware made up 6% of all beach litter collected in the 25-year anniversary tally (2011) of the International Coastal Cleanups (10.1 million pieces) [22]. How can you avoid this problem? You needn't bring your best silver, but taking your everyday dining room cutlery would go a long way to ensuring fewer plastic utensils on the beach – whether you are bringing your own food or going the takeout route. And more ecologically-friendly alternatives are available for all these items: compostable plates and bowls made of starch, recyclable plates fashioned from pressed or stitched-together leaves, and cutlery made of wood or from baked sorghum flour. The Internet is full of such alternatives, and with just a little research, you'll be sure to find biodegradable equivalents in a store near you.

Fig. 4.92 Spray bottles might be perfectly reusable – if refill liquids were available separately. Mediterranean, Turkey

Fig. 4.93 and Fig. 4.94 The mechanical parts of spray bottles – mostly plastic but often also containing metal parts such as springs – are built to take squeezing punishment in the cleaning frenzy. Mediterranean, Turkey.
Perfectly good, "high-tech" mechanical nozzles – here minus the siphon tube – are discarded by the millions, making for robust marine debris long after the container itself has deteriorated. Mediterranean, Turkey

Fig. 4.95 Sealant tubes: you'll find them in every home and every harbor and dock … and on most beaches. Mediterranean, Turkey

Fig. 4.96 You'll never get the last portion out with your "silicone" or "caulk" gun. Clogged nozzles also mean that such tubes – with their outrageously durable contents – are often discarded half full. Mediterranean, Turkey

Fig. 4.97 Even worn and without lettering or logos, the plastic bottle shapes of most foods and cleansers, lotions, and potions can tell beach detectives what they originally contained. Make a guess, and then take a walk down your local supermarket aisle and see if you can find the match. Mediterranean, Turkey

Fig. 4.98 Cute "personal-portion" plastic containers. The culprits are often nearby hotels that stock their guestrooms with single-use beauty care products rather than refillable dispensers. Take them home with you to avoid the wholesale disposal of half-filled products. Mediterranean, Turkey

Fig. 4.99 Plastic eventually bleaches and becomes brittle. One good step can reduce a household product to … microplastic. Mediterranean, Turkey

Fig. 4.100 Even unwieldy products, like this hose, ultimately break down into bite-sized pieces for marine wildlife. Caribbean, Guadeloupe

Fig. 4.101 Pens are a classic example of multicomponent marine debris: various plastics, metal springs, and ink. Few folks can resist taking a few free pens from the vacation hotel lobby. Mediterranean, Turkey

Fig. 4.102 The plastic body and bands of cheaper watches always survive much longer than the metal watch itself. Mediterranean, Turkey

Fig. 4.103 "Cutter knives" should be equally feared on beaches as on planes. The salty environment will hopefully dispatch the metal blade before you step on it. Insert a new blade: this is a reusable tool! Mediterranean, Turkey

Fig. 4.104 Paint brushes are used extensively in boatyards and harbors, making them common marine debris items. After all, no one likes to clean brushes steeped in paints and varnishes with evil-smelling, noxious fluids? Metal band remnants still visible on left. Mediterranean, Turkey

Fig. 4.105 The metal goes first; the bristles become microplastic, leaving the sturdy handles. The tubeworms on this one mean lots of time spent floating as marine debris at sea. Mediterranean, Turkey

Fig. 4.106 and Fig. 4.107 Clothes clips: traditionally made from a single piece of wood, now the domain of clever designers in plastics. Mediterranean, Turkey.
You'll find the full range of decomposition stages on most beaches. As a consumer, you might consider going back to one-piece wooden clothespins. Mediterranean, Turkey

Fig. 4.108 Even the simplest household items are often composed of several materials. Most clothes clip models still rely on metal parts. Mediterranean, Turkey

Fig. 4.109 Beachgoers stepping on inconspicuous, aged beach litter like this clothes clip may actually be important agents in breaking litter down (…into microplastic). Mediterranean, Turkey

Fig. 4.110 Make it a contest: see who can collect more in 10 min. Or each family member gets their own category of common beach litter. "The winner takes all" (home, hopefully). Mediterranean, Turkey

Fig. 4.111 Stripped of their shiny coating (and music), CDs are simply yet another type of plastic marine debris. Mediterranean, Turkey

Fig. 4.112 No, it's not washed up seagrass but probably 60 min of entertainment – the contents of a music cassette. Mediterranean, Italy

Fig. 4.113 And for you youngsters, the venerable compact music cassette consists of many small parts – reels, guides, and pads – including tiny metal bits. The magnetically coated tape (see above) is longer than a football field! Mediterranean, Turkey

Fig. 4.114 And for every music cassette, there is a plastic cassette box somewhere. Mediterranean, Turkey

Fig. 4.115 Portable coolers are rarely missing from a serious day at the beach, and chances are good they find their way back home in the evening. The same can't be said for damaged ice packs. Mediterranean, Turkey

Fig. 4.116 And while we're on the subject of beach picnics, the most direct household-plastic-to-beach-litter route is via the picnic basket. Always on the top 10 items collected during cleanups. Be conscientious – take home your used plates and cups. Even, better, check out the Internet for the many eco-friendly alternatives. Pacific, Mexico

Figs. 4.117 and 4.118 Plastic cutlery: a mainstay of every beach picnic. Try and replace them with biodegradable alternatives such as wood. Atlantic, USA

Fig. 4.119 Instantaneously recognizable around the world: the tiny spoons that automatically come with most takeout ice creams. Mediterranean, Greece

Fig. 4.120 Different cultures, different critically important decorative items: this fake grass is a common in sushi meals, ostensibly to help separate the fish from the condiments. Pacific, Japan

4.6 Plastic Bags and Other Packaging

Plastic Bags Hardly a day goes by when you and I are not handed a plastic bag with our purchases. Worldwide, nearly *two million* single-use plastic bags are used *every minute.* That's one trillion per year! The average EU citizen uses about 200 bags each year; in the USA that value is nearly doubled (one per day). Yes, they are practical and, depending on the luxury level of the shop, sometimes even elegant and artful.

As marine debris, however, plastic bags number among the top 10 items in annual beach cleanups every year. The 25-year anniversary international tally (in 2011) was 7.8 million or 5% of all items collected [22]. They are also on the top 10 list of dangerous debris items [5] because they pose a major threat to marine life. Perhaps the best documented problem is ingestion by sea turtles. The traditional explanation is that they confuse the wobbling bags in the water with jellyfish and start chomping on them. Maybe. Perhaps the turtles are simply targeting the organisms growing on the bags. Of course, from the evolutionary perspective, virtually everything that floats in the water is edible and worth a taste test. Unfortunately, the turtle throat is designed to make sure slippery prey can't escape, so once it's inside it can go only one way. And bags typically clog the digestive tract because, as you might guess, turtles didn't evolve to digest plastic. The result is slow starvation. Importantly, plastic bags pose a threat not only to wildlife but to humans as well. At sea, many an outboard motor has been destroyed by a snagged plastic bag that blocks the cooling water intake. This can cause an engine to self-destruct within seconds, leaving you adrift – and you know it will happen at the most inopportune time (i.e., storm coming up, darkness approaching, nothing left to drink, etc.).

But we simply can't leave a store without a plastic bag, right? As someone famous once said in another context, "Yes, we can!" Certain cities and even countries have forbidden the use of plastic bags. Bangladesh, for example, became the first country to ban plastic bags in 2002 after major flooding was attributed to bags blocking sewer systems. Several countries have followed suit, with others introducing taxes on each bag or fines for people using or selling them. In the USA, where an estimated 100 billion plastic bags are discarded every year, several cities and states have banned plastic single-use bags, most recently California. What can you do personally? The simplest action is to take your own cloth bag with you when shopping: it's reusable for years. If you do need a plastic bag, save it, and use it for the next shopping spree, for the garbage, or anything that promotes as many of the 6 "R"s (rethink, refuse, reduce, reuse, recycle, repair) as possible.

Packaging Not only is your purchase typically put into a plastic bag, but the product itself will almost always be sturdily packaged, usually in plastic, not to mention the product itself often enough being plastic. "Plastic in plastic in plastic"! Plastic packaging simplifies shipping and storage and keeps the items clean. It is also increasingly designed to be tamper-resistant, i.e., frustrating and nearly impossible to open quickly without a knife and without the real threat of mutilating the product and/or your fingers. All this means lots and lots of cleverly designed, sturdy packaging that – once opened – is quickly discarded. For almost every product, an abandoned package is lurking out there somewhere. Sometimes you'll find more plastic in the packaging than in the product itself. Lots of it ends up on the beach indirectly due to improper disposal or wind-blown transportation. Some of it is also instantaneous beach litter, i.e., when visitors unpack items directly on the beach. And what do folks unpack on the beach? Everything from new clothes to new sports equipment and food. Packaging poses the same threats as plastic bags, endangering sea life and boaters, not to mention the suffocation risk to small children. The key "R" here is "refuse": simply unpack your purchase at the store, and leave the packaging for proper disposal there – or avoid prepackaged items and foods in the first place.

Fig. 4.121 Plastic bags: high up on the top 10 list of marine debris items collected during beach cleanups and also topping the top 10 list of dangerous marine debris items. Mediterranean, Turkey

Fig. 4.122 Often what you see is only the "tip of the iceberg" … Pacific, South Korea

Fig. 4.123 … and must be pulled out of the sand for beach detectives to identify the source. This one is from a grocery store offering quick delivery and, praiseworthy, a refund for return of the bag. Pacific, South Korea

Plastic

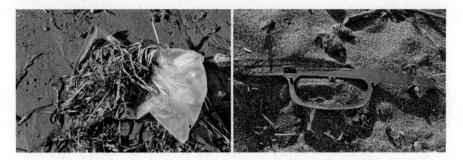

Fig. 4.124 Reinforced zip-locks, handles and grips survive the longest and sometimes provide the only clues for long-decomposed bags. Mediterranean, Turkey

Fig. 4.125 Ugly, illegal, commonplace – dutifully putting garbage into plastic bags and leaving them on the beach. Cultural tradition? Poor environmental awareness? Pure laziness? All probably play a role. Mediterranean, Turkey

Fig. 4.126 The great escape: thanks to wind, waves, dogs, and birds, wastes seldom remain in plastic bags for long. Only a small step from beach litter to marine debris. Mediterranean, Turkey

Fig. 4.127 and Fig. 4.128 Many bags (and other items) left on the beach end up as unsightly ornamentation draping nearby vegetation. In some places, folks refer to such snagged plastic as the "national flower." Red Sea, Egypt.
Some shoreline fences became so clogged with windswept plastic packaging that they billow like a sail and topple. Red Sea, Jordan

Fig. 4.129 This packaging originally contained swim goggles: instructions and plastic insert inside make this an easy one for budding beach detectives. More packaging than product! Mediterranean, Turkey

Fig. 4.130 Many items like this mask and snorkel set are first unpacked at the beach. Longer packaging? Flippers! Longer still? How about beach tents or umbrellas. Shop owners will hardly deny you leaving it all at the shop for proper disposal. Mediterranean, Turkey

Fig. 4.131 Woven plastic bags or sacs are particularly insidious ... Mediterranean, Turkey

Fig. 4.132 ... because they gradually unravel into thousands of individual strips of marine debris. Pacific, USA

Fig. 4.133 Woven plastic sacs might originally contain 25 or more kilograms of salt, rice, wheat, fertilizer, or the like, sometimes with an additional large internal plastic bag like here. When discarded and sand-filled, they are painstaking to collect during beach cleanups: watch your back! Pacific, USA

Fig. 4.134 Strapping bands are among the most indestructible of all packaging items. And also on the top 10 list of dangerous marine debris items. The metal clasps are the weak link. Mediterranean, Turkey

Fig. 4.135 Considering few people cook eggs on the beach, you'd be surprised how often egg cartons turn up at the water's edge. Enter "egg cartons and upcycling" into your computer's search engine for untold arts-and-crafts ideas! Pacific, South Korea

Fig. 4.136 For every fragile product, there is a cleverly designed, often plastic protective packaging. See Chap. 5 for more insights into packaging. Mediterranean, Turkey

Fig. 4.137 Plastic food packaging lands on the beach in all sizes. Here, a large 1 kg container for a cottage-cheese-like milk product. Mediterranean, Turkey

Fig. 4.138 The other size extreme? Packaging for a two-teaspoon-sized portion of yoghurt. Which "R" is best to apply here? Reject! Mediterranean, Turkey

Fig. 4.139 Plastic food wrappers for candy bars, ice cream, etc. – always near the top of the "dirty-dozen" lists of marine debris. Yes, they are usually sticky and sandy. Fight your inner couch potato, and take them home. Mediterranean, Turkey

Fig. 4.140 Chocolate spread, mustard, and many other foods – not only toothpaste – come in sturdy plastic tubes. You can never get the last drop out. Mediterranean, Turkey

Fig. 4.141 Condiments are a necessary ingredient for any beach picnic and also freely available at beachfront eateries, making them common beach litter items. They are never empty. Mediterranean, Turkey

Fig. 4.142 Quick, cheap, and practical: this fast-food philosophy means more packaging for the same content than larger, refillable dispensers. Like this single-portion ketchup packet. Mediterranean, Turkey

Fig. 4.143 The trend in many seaside hotels and restaurants is toward tiny, individual-portion trays, for your breakfast butter or marmalade, for example. Mediterranean, Turkey

Fig. 4.144 And, for the grand finale, what restaurant these days doesn't provide individually packed (in plastic) hand cleaning towels? Many folks grab a handful for future use in the car or on the beach. Mediterranean, Turkey

Fig. 4.145 With a little effort, snacking and clean beaches are compatible. It's "litterally" up to you!

4.7 Shotgun Shells

Anthropologists tell us we are basically hunters and gatherers. Modern beachcombers know all about the gathering part. But hunting? Yes, we have apparently also evolved to hunt down almost every creature we encounter. That includes animals that pose any kind of threat to us, which can mean that they are either big or have teeth, compete with us for food by eating crops, are too loud, smell as bad as we do in our un-deodorized state, or are deemed as pests because they can dig up or otherwise dirty our property. And, of course, most animals are a delicacy for someone somewhere. That doesn't leave a whole lot of fauna that we are benevolent toward. And due to the increasing decimation of animal adversaries that fall into any of the above categories, the tendency is to shoot at anything remaining that moves.

One of the reasons most of us so strive to protect beaches is because they may just represent the only natural landscapes we have ever encountered. Natural or seminatural landscapes, of course, mean a more or less functioning ecosystem, and a functioning ecosystem obviously must have plants and, yes, animals. In our strange obsession to keep at bay the poor animals that forage in barren sand dunes and beach habitats, there are those among us who use shorelines as hunting grounds. And what better way to control a wide range of pesky wildlife than with a shotgun? That explains one source of shotgun shells as beach litter.

A second source of such shells and their components is skeet shooting (at clay "pigeons") on cruise vessels. This onboard activity has decreased in recent times because it means that a lot of lead or other metal pellets, plastic wadding, and ejected shells end up in the sea – clearly a no-no today. Shooting is also very loud and clearly contravenes a main reason for going on a cruise in the first place. Finally, in times like these, who wants to hand passengers loaded weapons!

All this helps explain the abundant shotgun shells on many beaches of the world. As you can imagine, anything that is loaded into a gun and detonated – discharging lead or other metals faster than the speed of sound – is bound to be robust indeed. And robust in the barrel means robust as marine debris or beach litter.

Fig. 4.146 Shotgun shell wad and shot cup: formerly cardboard, cork, or felt, now high-density thermoplastic and long-lived. Protects barrel and prevents pellet deformation. Sardinia, Italy

Fig. 4.147 Telltale pellet indentations in the shot cup (viewed from above), which is fired out the barrel along with the pellets and makes hardy marine debris. Sardinia, Italy

Fig. 4.148 The tubeworms on the inside of this shotgun wad show that it spent time floating at sea. Stains of pellets still visible.

Fig. 4.149 Crimped business end of a shotgun shell. The cartridges and wads stem from shoreline hunters and skeet shooting on cruise liners. Mediterranean, Turkey

Fig. 4.150 Metal head (brass head) partially rusted away. Longitudinal ribbing still visible, crimped open end flaying. What beach creatures are being shot? Often anything that moves. Mediterranean, Turkey

Fig. 4.151 More advanced stage of cartridge decomposition: frayed case and rust marks of former metal head. Mediterranean, Turkey

Fig. 4.152 Being stepped on accelerates the decomposition of almost every beach litter type. One small step for you, one major leap for microplastic. Mediterranean, Turkey

Fig. 4.153 How many can you collect in 10 min? More than the number of lighters, clothes pins, or bottle caps? Mediterranean, Turkey

5

Foamed Plastic (Styrofoam)

Styrofoam, what a wonderful product: lightweight, waterproof, shock-resistant, long-lived, and harmless when in contact with food. This makes it useful for countless applications from food packaging to the impact-absorbing inner liners of motorcycle helmets. Its marvelous thermal insulating properties also make it a great container to keep cold drinks cool and hot food warm. And if your fingers are also important to you, styrofoam cups can help prevent scalding them while clutching your hot coffee-to-go. This means styrofoam protects three very important things in most people's lives – their brains, fingers, and their food, although not necessarily in that order: some folks attach greater importance to keeping their coffee warm than wearing a helmet.

Strictly speaking, styrofoam is actually a type of plastic, namely polystyrene. The product abbreviation stamped on most products – PS – refers to foamed polystyrene (see Fig. 4.1). Technically one can differentiate between extruded polystyrene (trademarked, styrofoam with a capital "S") and expanded polystyrene foam (EPS), much like James Bond distinguished between shaken and stirred martinis. For our purposes, we can stick to the generic term styrofoam (with a lowercase "s"), polystyrene, or just simply foam. The most important feature of this material as far as marine debris is concerned is that it consists almost entirely of air. But just like saying our bodies consist mostly of water, this doesn't really reveal much about the overall picture. What it does reveal is that styrofoam floats pretty much better than anything else, which makes it the preferred main ingredient in most products thrown to people who have fallen into the water (life savers, life buoys, life rings, or, if you want to sell them more expensively, "personal flotation devices"). This floatability means you'll find foam products and their remains on every beach in the world.

Foam or styrofoam has often been listed as a separate category on beach cleanup data sheets. This is because of its abundance and recognizability as the main or sole ingredient in many products. What are some of the most common foam items on beaches?

- Packing material, for one. These small S-shaped, spherical, peanut- or chip-shaped elements (technically referred to as "expanded polystyrene loose fill" but more commonly as foam peanuts, packing peanuts, packing noodles) are soft, pliable, and produced in untold numbers to softly bed sensitive products in boxes ("loose-fill packaging") and prevent them from being damaged during transport. How do they land on beaches? Mostly via sewage overflows, which shouldn't be too comforting a thought if you plan on taking a swim where the sewage system isn't functioning properly. A common, larger, and much more rugged styrofoam packaging type is the often very odd-shaped, white "frames" enclosing new computers, TV screens, kitchen appliances, and the like in their shipping boxes.
- Buoys and floats are a second very common styrofoam product. These spherical, cylindrical, lens- or spindle-shaped floats are made to take abuse. Their job is to mark the position of crab pots, to suspend oyster cultures, or to correctly orient and stretch out all manner of fishing nets (motto: lead weights along the bottom, styrofoam along the top). Although designed to take a beating, nothing defies the forces of the sea for long. The attached ropes or netting eventually cut through the buoys, tearing them free and landing them on beaches near and far.
- A final, major category of styrofoam-based marine debris and beach litter is food and beverage containers. This includes cups, bowls, plates and trays, takeout ("clamshell") packaging, egg cartons, etc. Takeout, single-use, disposable. Three strikes and you should be out! US standard drinking cup sizes include 6, 8, 12, 16, or 36 ounces, while elsewhere where the metric system applies, volumes might be 0.33, 0.5, or 1 liter, although who knows what the supersizing of our beverages and "jumbo-ization" of our meals will bring in the future. Cups and plates are always in the "top 10" debris items collected during beach cleanups, and many of these are made of polystyrene. And because product branding is the key to success in any business, such food and beverage containers are typically amply

furnished with names and logos that allow them to be traced to specific brands/food chains/franchises/cruise ships. This helps make life easy for beachcombing sleuths. How does this stuff become marine debris? Some is dumped by ships at sea, some enters via sewage overflows, but much is simply left on the beach by beachgoers. Sometimes exact addresses and telephone numbers on these items help track them to specific outlets. Often such detective work is child's play. On beaches with heavy tourism, for example, the location of discarded containers often corresponds to the nearest fast-food establishment along the boardwalk or beachfront: many items are discarded (or blown away) a stone's throw from their site of purchase.

In mentioning motorcycle helmets, I may have overstated the indestructible nature of styrofoam. Actually, while the material itself can be made quite durable (building materials, pipe insulation, buoys), many forms are fragile, breakable, and easily damaged. Light, heat, and chemical solvents can quickly destroy or dissolve even the best styrofoam, which is why you need gentler types of glues and paints for your polystyrene-based arts-and-crafts projects. The white styrofoam you are probably most familiar with is composed of tightly packed yet individually recognizable, expanded polystyrene beads. This is the type that scrunches and squeaks when you bend, twist, or put a knife to it. As styrofoam breaks apart, these little beads – every one of which also exhibits otherworldly electrostatic charge and sticking power – are gradually released. These are the final stages typically found on beaches. And if you can't immediately spot them, chalk it up to the wind. Styrofoam is super light and at the mercy of every breeze. Winds often come in from the sea (onshore winds), so check out the vegetation at the upper side of the beach (back shore): that's where this material tends to be, accumulated in depressions, lodged under, or snagged in plants. Offshore winds, in contrast, waft beached styrofoam back into the sea.

Hazards: As marine debris, styrofoam is long-lived. One estimate for the life span of a styrofoam cup is 50 years [1]. Floating at sea, small foam fragments pose an ingestion threat to wildlife. In addition, torn fishing net sections with intact styrofoam buoys may remain afloat over long periods and continue to capture and entangle marine organisms (so-called ghost nets). Like all plastics, styrofoam also leaches out chemicals added during the production process, pollutants that find their way into the food chain (and you know who stands at the end of this chain…) [2]. During beach cleanups, all foam items should be collected: don't worry, you won't be asked to pick up the individual polystyrene beads (or plastic resin pellets). Most marine debris, including polystyrene, should never be burned on beaches: it releases dozens of toxic and potentially carcinogenic compounds when thrown into the picnic fire.

What are the alternatives? The best strategy would be to avoid styrofoam whenever possible. Some cities, such as New York, have attempted to prohibit the sale, possession, and distribution of single-use expanded polystyrene (EPS) foam. This phrasing makes it sound a bit like a dangerous illegal drug, which is not such a bad analogy as far as marine debris goes. While that ban was overturned, the food and restaurant industry has hopefully gotten the message and begun exploring alternative options. One recent blog heralds "San Francisco bans polystyrene foam" [3]! In the case of "packing noodles," these can be replaced by old newspaper and junk mail or organic products such as popcorn or products made of rice starch (thermoplastic starch or bioplastics): equally lightweight, equal in size and cushioning effect, equally inexpensive, and totally biodegradable! As far as buoys are concerned, hollow ones made of heavier plastic would be an alternative: such floats would sink when broken or otherwise fragmented. Some countries have introduced workshops to inform fishers and help mold government policy [4]. In the takeout food industry, an alternative cardboard "clamshell" (maybe made of recycled paper?) for your cheeseburger would also do the trick: even if improperly discarded, it poses less of a marine debris problem than styrofoam.

References

1. Marine debris awareness poster (2016) https://web.whoi.edu/seagrant/outreach-education/marine-debris
2. Tanaka K, Takada H et al (2013) Accumulation of plastic-derived chemicals in tissues of seabirds ingesting marine plastics. Mar Pollut Bull 69:219–222
3. Ocean Conservancy (2016) https://oceanconservancy.org/blog/2016/07/07/san-francisco-bans-polystyrene-foam/
4. Lee J, Hong S et al (2015) Finding solutions for the styrofoam debris problem through participatory workshops. Mar Pollut Bull 51:182–189 https://doi.org/10.1016/j.marpol.2014.08.008

Foamed Plastic (Styrofoam) 163

Fig. 5.1 and Fig. 5.2 This is your typical styrofoam item: a fragment with individually visible polystyrene beads. Atlantic, USA. In the surf zone, the polystyrene beads break off individually or in clusters, forming untold microplastic particles. One report lamented "more foam plastic than sea foam" on a remote shoreline. Mediterranean, Turkey

Fig. 5.3 Great insulation properties mean food packaging material. This is the ubiquitous "clamshell" variety. Does takeout mean throw-out? Cardboard versions are a more ecological alternative. Pacific, USA

Fig. 5.4 and Fig. 5.5 Some like it hot but turn a cold shoulder to the marine debris problem. Pacific, Hawaii. Beach detectives can often identify fragments based on logos. Estimated life span of a styrofoam cup in the environment: 50 years.

Fig. 5.6 and Fig. 5.7 Plastic pollution is the sign of the times. Trace items back by their advertising slogans, lettering, or other unique design features defining professional product branding. This one perfectly sums up the entire marine debris issue. Pacific, USA. The tubeworm on this cup fragment indicates time spent floating at sea. Pacific, USA

Foamed Plastic (Styrofoam) 165

Fig. 5.8 The "law of marine debris aggregation": where you find one, you're bound to find many. Two cups may reflect "twin cupping": using two to shield fingers from overly hot or cold drinks. Stacking them, used or unused, slows degradation. Mediterranean, Turkey

Fig. 5.9 Leaving them stacked and wrapped can add another few years to their marine debris lives. Mediterranean, Turkey

Fig. 5.10 Final stage of a styrofoam plate. Reinforced and molded edges survive the longest in any marine debris item. Look for starch-based, biodegradable alternatives. Pacific, USA

Fig. 5.11 Beanbags: they get ugly fast on beaches and "spill the beans" when put under pressure (rambunctious kids or their overweight parents). Mediterranean, Turkey

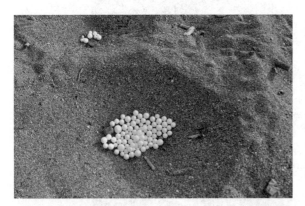

Fig. 5.12 The "beans" are millions of styrofoam beads or spherules, each beginning a new and long career as microplastic. Such beads, the same size as the infamous plastic resin pellets (Figs. 4.2, 4.4), tend to accumulate in every depression in the beach. One small hole is enough to start the flood. Mediterranean, Turkey

Fig. 5.13 Sometimes the beanbag "stuffing" contains other cheap foam fragments as well. Mediterranean, Turkey

Foamed Plastic (Styrofoam) 167

Fig. 5.14 Foam peanuts, packing peanuts, packing noodles: no matter what you call it, it's styrofoam and it's marine debris. Produced in billions to protect fragile items during shipping, there's only so much you can save and reuse at home or in the office after unpacking. Atlantic, USA

Fig. 5.15 S-, W-, figure-eight, or chip-shaped, typically white, green (partly recycled), or pink (antistatic chemicals added). Many biodegradable alternatives are available, including popcorn (motto: "high-tech to low-tech") or starch-based (bioplastic) products. Pacific, USA

Fig. 5.16 The bottom openings at both ends (to let water run off) reveal this to be a fishing box as used worldwide by commercial fishers. Mediterranean, Turkey

Fig. 5.17 About the only positive thing to say about vehicles on beaches is that they can crush marine debris and hasten its decomposition. Here, however, at the cost of producing microplastic avalanches. Mediterranean, Turkey

Fig. 5.18 Ultimately, even styrofoam gets ugly. Larger items break apart; the beads that make them up become discolored and shrivel. Reinforced rims tend to survive the longest. Mediterranean, Turkey

Fig. 5.19 Styrofoam's floatability makes it a common life buoy or "personal flotation device" material. Mediterranean, Turkey

Fig. 5.20 Floats and other buoys made of styrofoam are often "stabilized" with paint. Ropes and lines ultimately cut through most buoys, making each and every one ever sold into marine debris. Caribbean, Cuba

Fig. 5.21 Extreme lightweight and marvelous floatability makes styrofoam beloved in the fishing and sports industry. The larger the item, the more long-lived as marine debris. Pacific, South Korea

6

Hygiene

6.1 Personal Hygiene

Let's face it, we are human animals, and our bodies simply have to perform all the processes that life on this planet requires. We have to eat, and our metabolism therefore produces lots of fluids and solids that we try to distance ourselves from at the earliest possible convenience. And we almost daily need to manage the many features that characterize us as vertebrates and mammals, for example, our hair, skin, and nails. An impossibly confusing array of devices and instruments – not to mention creams, lotions, and potions – have been devised to this end.

The sad truth is that the cleaner, more aseptic, dapper and socially presentable we make ourselves, the dirtier and shabbier we tend to make our environment. Many of the devices and items we use to satisfy our hygienic, sanitary, and healthcare needs soon become, let's say, "icky": some after a single application (think cotton swabs or pads) and some at regular intervals (toothbrushes). Entire industries have arisen to provide us with "disposable," one-way products so that we don't have to clean or otherwise deal with "soiled" devices (think disposable shaving razors and razor heads). This means they are discarded in enormous amounts. The items in this category are usually instantly recognizable, and, because they typically consist of many different components (not just plastic), they warrant a separate heading here.

The fact remains that hygiene items are a common form of marine debris and beach litter. How do they get there? Well, most of us have a lot of free time on our hands when spending the day at the beach, and this provides a welcome opportunity to slowly and with relish (and in good light!) take care of some of the manicuring needs we tend to neglect during our otherwise hectic daily routine. Some of this paraphernalia is bound to be left on the beach (after all, who likes to take sticky, sandy, "empty," or otherwise partially consumed health products back home). And beaches are also a place where many "mishaps" that produce instant beach litter happen. Think of diapers as preprogramed to capture an accident just waiting to happen: most folks certainly don't want to take used diapers along for the long sweltering ride back home, right? And it's not our fault that the trash bins were already all full. (Check out the "organic debris" chapter for the type of stuff normally retained inside diapers.) Finally, don't forget that secluded beaches – or any beach at night – can be a very romantic place, ideal for a rendezvous, and you might guess the kinds of "personal hygiene" beach litter that intimacy can leave behind. Condoms, aka "Coney Island whitefish," are particularly abundant on the sand and in the water under boardwalks in said and other vacation hotspots.

All these considerations mean that you'll find many hygiene products in the broadest sense on your holiday beach. I use "in the broadest sense" here because, on the one hand, even the briefest of scans at your local supermarket or drugstore chain will reveal aisle upon aisle of personal hygiene and sanitary goods. On the other hand, there is a very "fluid" transition between hygiene and more hard-core medical wastes (which are treated separately in this book). This range of items and the potential hazards they entail are one of the reasons for wearing protective gloves during beach cleanups. You for good reason probably don't like to touch even your own discarded hygiene items, so picking up those left by other people with your bare hands is certainly not the best idea. Of course, if you're the type who thinks nothing of sharing a toothbrush, feel free to do a gloveless good deed.

Finally, personal care products and microplastics are also intimately related. Many lotions and cosmetics we apply to our skin contain tiny plastic "scrubber" particles. The same holds true for some toothpastes (up to 1.8% polyethylene (PE) microbeads by weight). All these products are eventually washed off. In the case of toothpaste, most – but never all – is spit out. Some of this material can pass through wastewater treatment plants and lands in our waterways. How much? One estimate is four million per day from each such facility in the USA. The total emissions into aquatic habitats from US wastewater treatment plants alone are estimated at eight *trillion* microbeads per *day* [1].

And what can you do in everyday life to go easier on the marine environment? How about replacing your single-use, disposable shaver with a more high-quality product or going electric? Biodegradable toothbrushes are available on the market, and some manual versions even feature replaceable or interchangeable heads: you don't have to throw away the whole thing when the bristles start to wilt. Does your deodorant have to come in a spray can with all its rugged and durable marine debris components? Diapers: the overall life cycle comparison between the disposable and the traditional washable cloth version comes out to a near stalemate as far as overall energy consumption is concerned [2]. The washable ones, however, certainly make for a lot less diaper-related marine debris. And on the beach, why not dare to go bare!

Fig. 6.1 and Fig. 6.2 Wind, waves, and hair. A problem that needs solving, making hair bands among the most common "personal hygiene" items lost (and found) in the water and on beaches (Mediterranean, Italy, Sardinia).
Hair clips also do the trick but are even more easily dislodged when cavorting in the waves or on the beach. Mediterranean, Italy

Fig. 6.3 and Fig. 6.4 A beach visit always involves multiple efforts to tame your hairdo. Mediterranean, Italy.
100 brushstrokes a day for lustrous hair? This hairbrush has met its match – well camouflaged, like many other aged and worn marine debris items on beaches. Mediterranean, Turkey

Fig. 6.5 and Fig. 6.6 This comb has outlived its usefulness and started a new career as marine debris. Caribbean, Cuba.
Carelessly discarded, brutally maltreated by the surf and merciless sun, but still hanging in there: an object lesson for the marine debris story. Pacific, USA

Fig. 6.7 Even the Romans used combs like this one. If all the teeth are very tightly spaced, chances are good it's a lice comb. Mediterranean, Turkey

Fig. 6.8 If we are not grooming our hair, then we are cutting or shaving it, more often than not with disposable products. Rusty razorblades make for unpleasant beach litter. Mediterranean, Turkey

Fig. 6.9 Some shaving accessories like this wooden shaving cream brush look downright quaint, but all marine debris takes on a different perspective when ingested by a dolphin or a sea turtle. Red Sea, Jordan

Fig. 6.10 and Fig. 6.11 Brushing twice a day will eventually do in even the best toothbrush. 435 were found in one recent international coastal cleanup. Mediterranean, Turkey. Some toothbrushes have apparently suffered an even crueler fate than being inserted into the human mouth. Mediterranean, Turkey

Fig. 6.12 and Fig. 6.13 What could this plastic stick have to do with hygiene? The indentations at the end are the giveaway clue for beach detectives. Pacific, USA. It's the spindle or shaft of a cotton swab (don't say "ear swab" says the doctor!). The indentations or cuts help hold the cotton. Mediterranean, Italy, Sardinia

Fig. 6.14 The cotton ends may be attached with glue and variously treated to maintain shape and prevent discoloration or mildew. The working ends may therefore be less "organic" than you think. Industry will need to return to cardboard shafts.

Fig. 6.15 And, of course, the "law of marine debris aggregation" applies to cotton swabs too: you'll almost never fine one alone. Compete with your friends: whether swabs, shotgun shells, clothes pins, lighters, or other common items – see who can collect more! Mediterranean, Italy

Fig. 6.16 As for most products, leaving them in their plastic packaging prolongs their life as marine debris. Mediterranean, Turkey

Fig. 6.17 and Fig. 6.18 The beach, a lovely setting and finally enough time and light to do your nails right. Atlantic, USA.
Products that protect your skin and lips are among the most common personal hygiene items left on beaches. Pacific, USA

Fig. 6.19 and Fig. 6.20 Tampon applicators are common marine litter personal hygiene products, here with "barrel" on top right and "plunger" on top left. Sometimes cardboard ("biodegradable and flushable," etc.), more often plastic. Mediterranean, Turkey.
The plunger pushes the tampon out through "petaled" end on left. Never flush the plastic type down the toilet!

Fig. 6.21 and Fig. 6.22 For those who like statistics: an average woman may use more than 10,000 tampons in her life. Is it any wonder that some of them end up on beaches? Tampons and fluids, including seawater, mean impressive size increases. A blue string is usually a good clue for beach detectives. Mediterranean, Turkey

Fig. 6.23 and Fig. 6.24 Oblong or "figure-eight" pads, with or without "wings," point to feminine hygiene products (sanitary napkins or pads) or, if larger, urinary incontinence pads ("adult nappies"). Mediterranean, Italy.
With some luck, you'll encounter only the packaging or peel-off part at the crime scene. Mediterranean, Turkey

Fig. 6.25 You can be sure that the diapers most parents with toddlers leave behind are not empty. Is this what they mean by "disposable"? Mediterranean, Turkey

Fig. 6.26 and Fig. 6.27 The adhesive mechanisms of diapers don't keep the beans from being spilled for long in the wind and waves (not to mention when inquisitive birds and dogs are around). Mediterranean, Turkey.
Diaper madness! The "law of marine debris aggregation"? A baby party gone horribly awry? Sometimes even hardened beach detectives are stumped. Mediterranean, Turkey

Fig. 6.28 If only we would try to keep our beaches as clean as our teeth. At least it looks like someone tried to extract the very last drop from this toothpaste tube. Caribbean, Cuba

Fig. 6.29 and Fig. 6.30 Any good beach detective will need only a few letters, part of a logo, or a color scheme to recognize most products around the world. Caribbean, Cuba. Product designers in non-English-speaking countries love to use English words and slogans. This often gives good beach sleuths decisive clues even about foreign products. Mediterranean, Turkey

Fig. 6.31 Using deodorant may be praiseworthy; there are more low-tech, environmentally friendly options. They instruct you not to throw it into the fire, but don't tell you not to leave it on the beach. Mediterranean, Italy

Fig. 6.32 Many a contact lens has been lost at the beach; the lens cases themselves perhaps less often. Although, come to think of it, blurry beach litter may well be less disconcerting. Mediterranean, Turkey

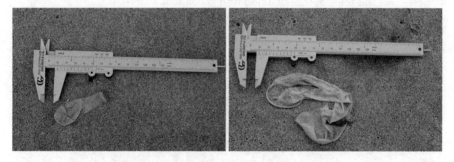

Figs. 6.33 and 6.34 Don't confuse balloons with their prophylactic cousins. Yet another reason to wear gloves during beach cleanups (caliper open to 1 cm)

Fig. 6.35 Condoms ("Coney Island whitefish") may be built to "take the heat" but are no match for beaches, which are even hotter and often wilder than sex. Atlantic, USA

Fig. 6.36 and Fig. 6.37 Like most marine debris items, the thicker parts survive the longest – like this telltale, rounded ring from a condom.
If it's more quadratic in outline, then you'll be relieved to know that it's probably simply a rubber band. Pacific, Japan

Fig. 6.38 and Fig. 6.39 Of course, for every condom there is a sturdy wrapper out there somewhere. In the heat of the battle, the marine debris perspective no doubt takes a back seat, as does the act itself sometimes too. Mediterranean, Turkey.
For the experienced beach detective, a few letters, a bit of color, or part of a logo is enough to give the decisive tip at the crime scene. Mediterranean, Turkey

6.2 Toilets and Co.

Billions of people on our planet mean hundreds of millions of "bathrooms" including all their furnishings. Is it any wonder that some of these items end up on our beaches? Most folks think that toilets and their contents are a very private affair. In fact, a discipline known as wastewater-based epidemiology is demonstrating that while people tend to lie, the urine they send down the drain doesn't. Wastewaters entering rivers tell researchers all about the level of drug use in the upstream city [3]. And people flush many other items down the toilet that they shouldn't. The most distinctive bathroom-related item on beaches? The toilet itself (Fig. 6.40). Globally seen, toilets can range from the squatting kind (flat enamel basin with a central hole and elevated footrests left and right) to the sitting type with seat and cover. The high end is cornered by electronic, heated wonders with more buttons than your TV remote to help cover up undesired sounds and odors and wash your butt afterward. The uninitiated would be well advised to receive instructions before visiting any type.

Most readers will probably be familiar with the sit-down, flush toilet. This includes the bowl itself, the cistern and lid to hold the flushing water, and the toilet seat and seat cover, along with many levers, pipes, floats, and other sundry fittings. Where there are toilets, you'll find all the related items: toilet brushes, brush holders, toilet bowl cleaner and disinfectant containers, toilet bowl tabs (and their holders), air freshener sprays, etc. And, for when things go wrong in bathroom or kitchen, plungers.

Interestingly enough, most of these sanitary articles are common marine debris items. That includes entire toilets. Think I'm kidding? The 2017 international beach cleanup yielded 56 of them [4]. How do most toilet articles land in the water or on the beach? Well, toilet bowls themselves are no doubt "land-based," i.e., someone drove by and disposed of them directly on the beach. Most other toilet-related items are probably washed into the sea due to improper waste disposal and wastewater management. Others are dumped offshore by commercial vessels or thrown overboard from excursion or leisure boats. Beachside public restrooms are another likely source. Many beaches, however, are not equipped to meet visitors' every need, or shall we say even their most basic need. This calls for self-sufficiency. If the calling comes, you don't want to be surprised by the lack or dreadful condition of the sanitary facilities. Most folks would rather be on the safe side and at least bring their own toilet paper. Ever heard of someone taking used toilet paper back home…?

Be that as it may, all toilet-related objects should be collected during beach cleanups. Although the sun and saltwater tend to decontaminate and make debris less hazardous, toilet articles and personal hygiene items are one reason why wearing gloves or using a litter pickup stick is recommended. And don't forget, it could be worse elsewhere: the UN has declared World Toilet Day (November 19) for the two billion or so people who don't have access to toilets….

Fig. 6.40 Don't ask, don't tell.... OK, this one is posed, but it was moved only a short distance from the dump of a beachside camping site. Mediterranean, Turkey

Fig. 6.41 You can hear the ocean roar without flushing! Wooden toilet seats, whether painted or laminated like this one, float quite well. Mediterranean, Italy

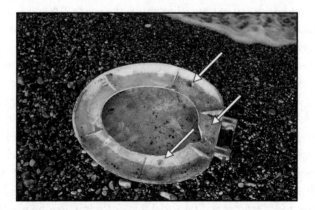

Fig. 6.42 Smaller, cheaper, and flimsier plastic seat of the type found on many boats. Encrusting organisms (arrows) mean time spent floating at sea. Mediterranean, Turkey

Fig. 6.43 A toilet without plumbing is … a mess. Mediterranean, Turkey

Fig. 6.44 Where there are toilets, there will be toilet brushes. Mediterranean, Turkey

Fig. 6.45 The working end of this brush is eroded away, hopefully in the surf zone and not through recent regular use. Some models have replaceable brush heads to extend useful life and reduce plastic waste. Mediterranean, Turkey

Fig. 6.46 And where there are toilet bowl brushes, there must be brush holders. Mediterranean, Turkey

Fig. 6.47 If the going gets tough, you'll need a plunger. This one apparently met its match. Mediterranean, Turkey

Fig. 6.48 For beach detectives, the "swan neck" is a dead giveaway for toilet cleaner. Lets you reach those hard-to-get-at, under-the-toilet-bowl-rim areas. Mediterranean, Turkey

Hygiene 187

Fig. 6.49 and Fig. 6.50 And don't forget toilet bowl sanitizing and aroma tab holders: built to take rigorous flushing punishment, making them durable and feeling right at home in the surf too. Mediterranean, Turkey.
Hinged toilet bowl tab holder for easy tab exchange, meaning at least two pieces of marine debris. Mediterranean, Turkey

Fig. 6.51 Toilet paper tube. Taking toilet paper to the beach might be a mark of self-sufficiency, but you can't help wonder where all the paper has gone. Check out the Internet: alternative, "tubeless" toilet paper rolls are available! Pacific, Japan

Fig. 6.52 A showerhead: you could probably put together a makeshift bathroom with all the marine bathroom-related marine debris items you'll find. Mediterranean, Turkey

Fig. 6.53 Just when you think you've seen it all. Marine debris: really everything including the bathroom sink! (for the kitchen sink, see Fig. 3.49). Caribbean, Guadeloupe

Fig. 6.54 Sometimes a clue like this built-in soap tray is necessary to give beach detectives that "sink"ing feeling. Caribbean, Grenada

References

1. Bråte ILN, Blázquez M et al (2018) Weathering impacts the uptake of polyethelene microparticles from toothpaste in Mediterranean mussels (*M. galloprovincialis*). Sci Total Environ 626:1310–1318. https://doi.org/10.1016/j.scitotenv.2018.01.141
2. Bast W-A, Diehl S (1991) Produktlinienanalyse Babywindeln: eine vergleichende Untersuchung von Baumwoll- und Höschenwindeln. Knoll, Klein-Umstadt, p 152
3. Editorial (2016) Bowled over: assessing the contents of the toilet bowl in the name of crime prevention. Nature 537:280. https://doi.org/10.1038/537280a
4. Ocean Conservancy (2017) https://oceanconservancy.org/wp-content/uploads/2017/06/International-Coastal-Cleanup_2017-Report.pdf

7
Medical Wastes

There is a lot of unsavory marine debris out there. But medical wastes clearly add an additional, downright sinister dimension. While some of them may be loosely related to the personal hygiene items treated in the previous chapter, the objects in this category are typically one step more hardcore, the ingredients at least one grade more powerful, and the infection risk from contaminated parts and contents an order of magnitude higher. And it doesn't need to have a hazard symbol stamped on it: no matter what the item may be, it can spell no good. Are sick people medicating themselves on the beach? Is the beach being used by drug addicts? Is the municipal waste treatment system down for repair? Has some company improperly disposed of (i.e., dumped into the sea) medical or research lab waste? These are some of the potential scenarios explaining your cleanup predicament.

The amount of medical waste produced in the USA alone is estimated at 3.2 million tons or about 2% of the total municipal solid waste stream [1]. Medical wastes encompass a huge range of items: thousands of different human diseases, ailments, and injury types mean tens of thousands of different pills, ointments, bandaging, syringes, application devices, surgical aids, and postsurgical hospital supplies with all their containers, adapters, connectors, valves, and tubing. This includes all infectious waste, hazardous (including low-level radioactive) wastes, and any other wastes generated from all types of healthcare institutions, including hospitals, clinics, doctors' (including dental and veterinary) offices, and medical laboratories [2]. Even if you are not a physician and can't put a name to it, chances are good you'll recognize the marine debris that belongs in this category.

The most common medical beach litter item is syringes – many thousands are collected every year during international coastal cleanups, and they are on the top 10 list of dangerous debris items [3]. "Hypodermic syringes," "hypodermic needles," or simply "needles" come in all shapes and sizes, all of which should put you on high alert. Whether reusable and made of glass and metal, or single-use, disposable, and made of plastic, the first thing you'll want to do is put on your shoes. And then your gloves. The many types are usually specifically designed for particular medical and scientific tasks, from insulin shots for diabetics to cortisone injections for insect stings, to injections of radioactive substances in research labs and hospitals. Sizes (capacities) generally range from 2.5 milliliters (ml) to 60 cubic centimeters (cc), but slender microsyringes with microliter (µl) capacities to jumbo syringes (250 ml) for dispensing larger volumes of liquids also exist. Important distinguishing features include length, diameter, and volume of the syringe body. The needle itself provides additional clues about the intended application (length, diameter, wall thickness, tip design).

How do syringes get on beaches? One sad direct source is drug addicts: intact, undamaged syringes in dune areas, below a boardwalk, or near a beach access are hints at this source, especially if other drug-related items are also present. Such paraphernalia on and around the beach can include tiny plastic zippered bags (for crack cocaine). More eroded and abraded syringes may point to indirect sources such as via sewage systems or waterways entering the sea. Rarely, major medical waste washups occur when companies entrusted with properly disposing of research lab and hospital wastes simply dump them illegally at sea. This requires quickly informing the authorities and necessitates beach closures. In summer 1987, for example, beaches in New Jersey had to be closed twice after medical waste (apparently originally from hospitals in New York) began washing up on nearly 70 miles of coastline. The state lost an estimated 1 billion dollars in tourism revenues – not including the cleanup costs – and introduced stricter regulations [4, 5]. A similar incident affected beaches in Long Island, New Jersey, and Rhode Island in 1988 [1]. The sources were identified as a landfill (including barges transporting waste to it), marine transfer stations, combined sewer overflows, raw sewage discharges, storm water outlets, and illegal dumping. Ouch! Again, costs were estimated at nearly 1.5 billion dollars.

Back to syringes. They are composed of many parts – barrels (shafts), plungers, needles, and their protective caps – and you will often find them separately. It's important to be able to recognize them. Sunlight and sand abrasion will reduce the transparency of the barrel, gradually erasing any lettering and the scale (volume) markings. As you might guess, the stainless steel needles pose the greatest risk. They can inflict truly ominous puncture wounds. Residual substances in the hollow needles may be toxic in and of themselves or be contaminated by any one of the dozens of disease-causing agents that coursed through the veins of the original "patient." In some intravenous-drug-using communities, hypodermic needles may be contaminated with the AIDS virus. If you have the great misfortune of stepping on a needle, go straight to a local hospital or physician, even if no wound is immediately apparent. Take the syringe with you in case laboratory analyses are necessary. For your sake and that of other vacationers, you should collect syringes and their parts, along with any and all other medical wastes. During beach cleanups, wear shoes and wear good gloves; the coordinators or site captains should be informed of any such finds and will probably provide you with a special container (with strong walls and a lid) for any needle-bearing items. "I've seen the needle and the damage done," sang Neil Young.

References

1. Wagner KD (1990) Medical wastes and the beach wash-ups of 1988: issues and impacts. In: Shomura RS, Godfrey ML (eds), Proceedings of the Second International Conference in Marine Debris. U.S. Dep. Commer., NOAA Tech. Memo, NMFS, NOOAA-TM-NMFS-SWDSC-154. https://swfsc.noaa.gov/publications/TM/SWFSC/NOAA-TM-NMFS-SWFSC-154_P811.PDF
2. Lee M (1988) CRS report for congress: infectious waste and beach closings. Congressional Research Service, Library of Congress, Wash. D.C. 9 Sept. 1988
3. Ocean Conservancy. act.oceanconservancy.org/site/DocServer/MarineDebris.pdf?docID=4441
4. Marine Defenders. http://www.marinedefenders.com/marinedebrisfacts/impact.php
5. State of New Jersey (2016) Guidance document for Regulated Medical Waste (RMW). http://www.nj.gov/dep/dshw/rrtp/rmw.htm

Fig. 7.1 Although it may look harmless, this is one of the most insidious marine debris items. Can you guess? Syringes are by far the most common medically related litter items and can mean many things, none of which are good. Adriatic, Slovenia

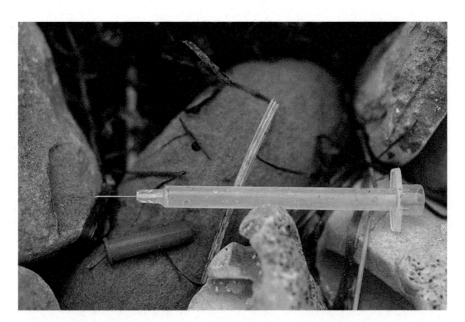

Fig. 7.2 Removing the red protective cap reveals the metal working end. The plunger is missing. You would really hate to step on this. Adriatic, Slovenia

Medical Wastes 195

Figs. 7.3 and 7.4 Although complete syringes are a high-tech symphony of plastic (or glass), metal, and rubber, the metal working end typically (thankfully) rusts away first. Plungers with rubber stopper visible inside shaft. Red Sea, Egypt

Fig. 7.5 Stepping on any aged and brittle marine debris hastens its decomposition. Mediterranean, Turkey

Figs. 7.6 and 7.7 Learn to recognize individual syringe parts, like these plungers, with or without the rubber plug. Caution: the "law of marine debris aggregation" means if you see one, there are bound to be more nearby. Mediterranean, Turkey

Fig. 7.8 A tight-fitting, rounded bottom end (on right) replaces the rubber stopper function in some syringes. Mediterranean, Turkey

Fig. 7.9 Another telltale item is the plastic needle cap, nestled up against styrofoam here. For every cap, there is a needle somewhere. Mediterranean, Turkey

Figs. 7.10 and 7.11 Sometimes you'll find the cap – plus needle – detached from the syringe. Mediterranean, Turkey

Fig. 7.12 Medications. We all take them. Sometimes even at the beach. Elongated bubbles mean capsules, and crunched bubbles mean they have been taken. Mediterranean, Turkey

Fig. 7.13 These bubbles haven't been pressed, and the decomposing pills are hazardous wastes. Pacific, South Korea

Fig. 7.14 Medication packaging. Their mix of plastic, metal, adhesives, and tamper-resistant design makes for a durable product – and long-lived marine debris. Round bubbles mean tablets. Mediterranean, Turkey

Fig. 7.15 At beach cleanups, you won't want to open closed medication packaging. Wear gloves, and treat all medical wastes as hazardous. Mediterranean, Turkey

Fig. 7.16 Plastic caps of medications (here embossed with company name and logo to help beach detectives) often have a characteristic, spiraled lower part to secure tablets in their tube. Mediterranean, Turkey

Medical Wastes 199

Fig. 7.17 And this is where it goes from unpleasant or disgusting to criminal. Packaging with tubing, like this infusion apparatus, almost always points to improper, larger-scale disposal of medical wastes. Mediterranean, Italy

Fig. 7.18 When you encounter medical marine debris like this oxygen mask, keep a sharp lookout for similar items nearby ("law of marine debris aggregation"!), and inform lifeguards, beach patrols, or local police. Atlantic, USA

Fig. 7.19 Let's close this chapter with something a bit less disconcerting. Nonetheless, "adhesive bandages" are medical wastes. Summer vacations mean cuts and bruises, so they are in everyone's first aid kit. Still packaged, less concern. Pacific, USA

Fig. 7.20 More often than not, "Band-Aids" land on the beach because they don't really stick very long in the surf. Blood and other bodily fluids: a potential health hazard and more concern. If loose, take them off (and take them home). Atlantic, England

8
Furniture and Furnishings

8.1 Beach Furniture

We all love sandy beaches, right? It can't be because of the sand itself, though, because very few people apparently want full body contact with it for any length of time. In fact, you'll rarely see anyone sitting directly on the stuff. Most folks want a barrier, at least a towel or air mattress, preferably something that distances us even more such as picnic chairs and full-blown sunbeds. And let's not forget the furnishing excesses on beaches such as rugs, bean bags, and the like – all designed to keep that nasty sand at bay.

We don't seem to love the beach for the sun either: most visitors bring along umbrellas, beach tents, and other gangly, shade-providing contraptions to hide under. And who likes to eat and drink directly on the bare sand? Spread out a tarp and giant towel, or set up your folding camping table. If you are a beachside business, provide an umbrella and sunbed with small table and chances are you'll find them occupied all day. Besides, if you provide furniture, you can also charge for it. A few dollars rent for the umbrella and a few more for the sunbed. A few turnover times a day, hundreds over the season – this sounds like a profitable and healthy outdoor business! Of course, it's an incentive to put up as many sunbeds and umbrellas as possible. In some places, this means row upon uninterrupted row of densely spaced sunbeds and umbrellas from the first dry sand at the high-tide mark all the way back to the bar, restaurant, or parking lot.

In the marine debris context, this means lots of furniture on beaches. One recent international coastal cleanup yielded enough items to furnish an entire studio apartment, including an air conditioner, sink, refrigerator, dishwasher, oven, microwave, washing machine, couch, table and chairs, television set, coffee table, rug, curtains, toilet, dresser, desk, and a bed complete with mattress pillows and pillow cases [1]. Of course, most furniture doesn't hold up well against the natural elements – the sun, water, salt, and wind. You may not like rental sunbeds and mattresses, which tend to be icky from the sweat and sunscreen lotions of innumerable previous sunbathers. The solution, of course, is to bring your own. And what's the main criterion it must fulfill? Correct, it has to be lightweight. And cheap. Lightweight combined with cheaply built and foldable equals flimsy and fragile, which brings us to other natural elements, namely romping kids and overweight vacationers: they quickly create "compromised" equipment, which means that a lot of personal furniture remains on the beach. After all, who likes to take broken things back home, and they typically won't fit into the skimpy trash bins most public beaches provide.

Beachside businesses and vacationers are not the only sources of furniture on the beach. Some comes from the sea, for example, the ubiquitous plastic, stackable chairs on the decks of excursion boats and less expensive cruise ships or seaside terraces. It doesn't take much of a wind or too much of a swell to send these items skidding off into the water. Illegal dumping activity (off the cliff from a serpentine coastal road…) also contributes its fair share to furniture marine debris. Finally, folks who make the effort to lug all that furniture are usually also well equipped with other expedition gear and well stocked with provisions. This makes them profuse generators of every conceivable category of beach litter. The extensive array of furnishings, the many materials they are made of, and their role as garbage hotspots warrant a separate category in this field guide. Let's take a look at what's up on the furniture front and how it contributes to the beach litter and marine debris problem.

Fig. 8.1 Set up furniture and it will be occupied. And what do people do all day? Eat, drink, and smoke, making beach furniture a beach litter factory. Mediterranean, Turkey

Fig. 8.2 Sunbeds and umbrellas are often only a slight miscalculation away from becoming marine debris. Mediterranean, Turkey

Fig. 8.3 Beach furniture holds up poorly against the tough natural elements in the beach environment. Even minor damage means a pinched butt and instant rejection. Mediterranean, Turkey

Fig. 8.4 Other "natural" elements such as overweight visitors, rambunctious kids, or rowdy partyers give beach furniture the "coup de grâce". Mediterranean, Turkey

Fig. 8.5 A buckled or missing leg, broken armrest, or frayed seat edge makes for instant "mega beach litter". Caribbean, Guadeloupe

Fig. 8.6 Furniture "game over" – let the beach debris game begin. Mediterranean, Turkey

Fig. 8.7 and Fig. 8.8 You'll sometimes have to look twice to recognize furniture fragments. Rounded edges, reinforced corners, and legs survive the longest and provide valuable clues at the crime scene. Mediterranean, Turkey.
The final stage of plastic beach furniture: lone sunbed "legs". Mediterranean, Turkey

Fig. 8.9 More sophisticated and adjustable recliners with metal frames don't fare much better than their all-plastic cousins. Atlantic, USA

Fig. 8.10 The salty air quickly turns beach furniture into unappetizing beach litter. Mediterranean, Turkey

Fig. 8.11 The more metal, the uglier. Mediterranean, Slovenia

Fig. 8.12 Beach umbrellas become crime scenes in their own right – a cornucopia of beach litter. Mediterranean, Turkey

Fig. 8.13 One wind gust can make instant marine debris. Mediterranean, Turkey

Fig. 8.14 On sea turtle nesting beaches, sticking your own umbrella into the sand endangers the eggs and hatchlings. Mediterranean, Turkey

Figs. 8.15 and 8.16 Silent witnesses of epic struggles to make the harsh beach environment more cozy. Bent, buckled, snapped, rusty poles are typical remnants of beach umbrellas, often recognizable by their plastic adjusting and arresting mechanisms. Mediterranean, Ibiza

Fig. 8.17 Every "fixed" umbrella has its holder. This water-filled plastic model has already made the transition to beach litter. Cement holders make for very long-lived litter (but can help avoid harming baby sea turtles by people inserting poles directly into the sand during the nesting season). Mediterranean, Turkey

Furniture and Furnishings 209

Fig. 8.18 The more furniture, the higher the profits and the greater the potential for creating marine debris. Atlantic, England

Fig. 8.19 Time and tide quickly turn even the sturdiest furniture into marine debris. Mediterranean, Turkey

Fig. 8.20 Few sunbed mattresses stand the test of time. Mediterranean, Turkey

Fig. 8.21 Soggy bed mattresses present a beach cleanup challenge – watch your back! Mediterranean, Turkey

Fig. 8.22 Box spring mattresses, all the rage. And rage it causes as beach litter as well. Mediterranean, Turkey

Fig. 8.23 Beach furniture taken to horrible excess. Rugs to ensure your water sports equipment isn't scratched. And this is on a sea turtle nesting beach! Mediterranean, Turkey, Çalış/Çiftlik

Fig. 8.24 It may be the beach, but you'll have to make an effort to find the sand at this furniture crime scene. Mediterranean, Turkey

8.2 Appliances and Home Electronics

Today, most of us are surrounded at home and elsewhere by a vast array of devices ranging from TVs and computers to irons and refrigerators. Much of this is electric or electronic. Every person generates an estimated 7 kg (15+ pounds) of so-called e-waste per year! That's the average, with folks in the Americas producing 12 kg and those from Europe 16 kg per capita [2]. A few years ago, the total amount dumped reached a whopping 42 million tons per year globally, much of it a mix of kitchen, bathroom, and laundry equipment [3]. If you are surprised at this, how many times have you already replaced your computer? Or your cell phone? Why is this necessary, where do all these devices go, and why is it a marine debris issue?

You've no doubt heard of "planned obsolescence." Manufacturers, for example, introduce new "must-have" models at ever shorter intervals. In the best case, the new version is significantly improved over last year's; in the worst case, it's a mere stylistic change. More nefarious is when products are designed to fail early on by incorporating an often irreplaceable, low-quality part. Or the device cannot be opened and serviced, or it is designed for one-time use (think disposable camera), or a battery cannot be replaced.

Do you sometimes wonder where all these unfashionable or prematurely nonfunctional devices land? Some go to specialist recyclers, lots also "south" in the framework of illegal e-waste exports for children to salvage in dumps and scrap yards. In 2012, China processed about 70% of the world's e-waste, with much of the remainder going to India and other countries in eastern Asia and Africa, including Nigeria [4]. Despite comprehensive directives and laws, the EU and the USA and Canada dispose domestically of only 40% and 12%, respectively, of their e-wastes [5].

And some definitely lands in the sea and on beaches. The 2016 international coastal cleanup, for example, yielded 97 TVs and 28 refrigerators [6]. They can take the same pathways as most other debris: dumped over the cliff from a coastal road, legal and illegal dumps too close to the water, swept into the ocean after having been thrown into streams and canals. Some have been dumped into the sea directly along with other garbage. Many appliances have air pockets or insulation that can keep them afloat. Finally, some appliances are simply left behind on the beach (cheap, greasy barbecue grills, for example). Check out Chap. 3 (metal) for some bigger kitchen appliances.

Is this all merely an esthetic problem? No. Beyond containing valuable components (think precious metals in cell phones), many devices contain heavy metals and toxic chemicals. Personal computers contain mercury, arsenic, and chromium [3]. One newly recognized class of threat is polybrominated diphenyl ethers, widely used as flame retardants in electronics. Finally, keep an eye out for sharp, rusting edges of appliances, and watch your back while extracting partially buried refrigerators and the like from the sand during cleanups.

Suggested solutions to this pollution problem include an expanded formal global protocol, better domestic regulations, a stronger role for the "United Nations' Solving the E-waste Problem Initiative," and special payments to countries receiving such wastes for disposal – and enshrining consumers' responsibility in regulations (that's you and me!) [3]. Where are we now? The amount of dumped telephones, televisions, and appliances doubled between 2009 and 2014 [2], and that trend will not be stopping anytime soon. Expect to find more on the beach in the future.

Fig. 8.25 A recent international coastal cleanup yielded 97 TVs. What should you do if you find one on the beach? Mediterranean, Greece

Fig. 8.26 and Fig. 8.27 Before removing it, how about expanding your entertainment program by watching real live TV and saving electricity to boot? Mediterranean, Turkey.
With a little luck you'll find a remote control in equally good condition. Mediterranean, Turkey

Fig. 8.28 Piles of marine debris with unique "e-wastes" are practical orientation points (and apparently attractive places to linger - "let's meet at the garbage pile!") along endless beach expanses. Mediterranean, Greece

Fig. 8.29 and Fig. 8.30 You can't escape computers, even on the beach. Flat computer monitors have meant the demise of untold, often perfectly functional older models. Caribbean, Guadeloupe.
Once they become beach litter, they'll need more than any computer doctor can offer. Why install a romantic screensaver scene when you can have the real thing as a backdrop. Caribbean, Guadeloupe

Fig. 8.31 You don't think marine debris is funny? Then call and complain: you'll find company logos and addresses along with all contact information on many beach litter items! Mediterranean, Turkey

Figs. 8.32 and 8.33 OK, here are some handsets too; now nothing can stop you. Mediterranean, Turkey

Fig. 8.34 Some beach litter items are hard to explain. An "extreme ironing" competition (check out the Internet!) gone awry? A lethal marital spat? Mediterranean, Turkey

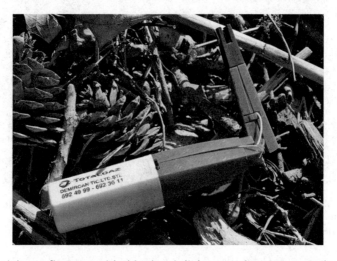

Fig. 8.35 Light my fire? Not with this electric lighter. Mediterranean, Turkey

Fig. 8.36 The selfie generation is also leaving its mark on the world's beaches. A deadly practice: taking selfies kills more people every year than the sharks lingering just off the beach. Taking selfies up close with wildlife is a major no-no. Five selfie sticks were found in a recent international coastal cleanup. Mediterranean, Turkey

References

1. Ocean Conservancy. https://oceanconservancy.org/wp-content/uploads/2017/04/2014-Ocean-Conservancy-ICC-Report.pdf
2. Trend watch (2015) E-waste estimates. Nature 520:413
3. Zhang B, Guan D (2016) Take responsibility for electronic-waste disposal. Nature 536:23–25. https://doi.org/10.1038/536023a
4. Zhang K, Schnoor JL, Zeng EY (2012) E-waste recycling: where does it go from here? Environ Sci Technol 46(20):10861–10867. https://doi.org/10.1021/es303166s
5. Baldé C, Wang F et al (2015) The global e-waste monitor 2014. Quantities, flows and resources. United Nations University, Tokyo and Bonn 90pp
6. Ocean Conservancy (2017) 30th Anniversary International Coastal Cleanup. https://oceanconservancy.org/wp-content/uploads/2017/04/2016-Ocean-Conservancy-ICC-Report.pdf

9

Apparel

9.1 Clothing

"Clothes make the man." And certainly the woman. If that's true, then fashion dictates that we must make and remake ourselves regularly. Even though going to the shore is all about discarding clothes, you'll still have to wear something to get to and from the beach. And most of us will probably be wanting to cover key body parts once we get there. Even if it's skimpy, though, it's always enough to make a fashion statement. To paraphrase Frank Zappa: "Everyone on this beach is wearing a uniform and don't kid yourself." And it's always enough to contribute to marine debris and beach litter.

Even if we might wish that some people would cover more of their body's biggest organ (that's the skin, folks), most do not. Whether unsightly or not, sun and skin make for a dermatologist's worst nightmare. Thankfully, the word has spread that liberal and repeated applications of strong sunscreen can help determine how you will look at an older age and whether you will reach such ripeness in the first place. Of course, the best combination on hot beaches is sunscreen plus some sort of loosely fitting clothing (and a beach umbrella, but that's another marine debris issue – see Chap. 8 (furniture)).

The number of clothing items on the beach is a matter of simple mathematics: human anatomy dictates that there will always be more items of clothing on the beach than persons, and not all of it goes home. Why? Maybe it didn't survive the beach volleyball game. Perhaps it was blown away unnoticed by the wind or was ripped off by an unexpected wave and couldn't be retrieved in the foamy water. Or fell victim to a digestive mishap (Montezuma's revenge) or became sticky, sandy, and yucky from melted ice cream or other food accidents. Like most any item dropped on the sand, it could also simply have become covered and been forgotten on the hasty retreat home in the low evening sun.

But most clothes are organic and can't be a significant marine debris problem, right? Wrong. In fact, most clothes are made of synthetic fibers these days, and even your "cotton" apparel may be variously fortified with fibers that don't grow on bushes or trees. That could be the elastic strip in your waistband, various shiny decorative elements, or the extra-strength thread used. Sometimes the lining will be cotton and the outer material synthetic. Or vice versa. And, of course, don't forget the sewn-in washing instructions, company logos, and other stiffish tags that are seldom organic. High-tech apparel, such as used for water sports and other outdoor activities, is rarely made of cotton. Finally, many clothing items are chemically treated or impregnated to withstand mold, either ward off or pass through moisture, or fulfil the many other crucial duties that discerning consumers demand these days.

What are the dimensions of this waste category? Nearly 10 billion kilograms of apparel end up in landfills every year, and making one pair of jeans apparently consumes between 3000 and 10,000 liters of water (enter "water footprint" into your computer's search engine) [1]. And the solutions? Some companies have gotten together and formed a Sustainable Apparel Coalition with their own sustainability scores (the Higg Index) [2]. Other code words include bio- or organic cotton, biodegradable clothing, eco-friendly tanneries, recycled rubber shoe soles, or 'from waste to wear' (e.g., making socks from recycled fishing nets [3]). A key issue these days is microplastics, and clothing is a major producer of synthetic microfibers, both via wastewater and the atmosphere [4]. Every dressing and undressing, and every washing, yields untold microplastic pieces that enter the environment and, as we now know, also the marine food chain and ultimately our own bodies.

Whether organic or synthetic, however, most apparel is made to be robust and withstand everyday wear and tear. All this means clothing survives for quite some time in the water and on beaches too. All of it should be removed during beach cleanups. And who knows, you might actually come across a valuable, near-pristine, fashionable item you can call your own!

Apparel   221

Fig. 9.1 Oilskins – work clothes worn on almost every yacht or boat by serious crew members who work on deck in nasty conditions. Made to take a beating means hardy marine debris. Gulf of California, Mexico

Fig. 9.2 Winter coats on tropical beaches may have come from faraway climates. If you're an adventurous beach detective (and wearing gloves), you can check the pockets for clues. Mediterranean, Croatia

Figs 9.3 and 9.4 Pulling protruding parts of coats and other clothes from the sand typically reveals their full glory (and hopefully no associated body parts). Pacific, South Korea

Fig. 9.5 Plastic zippers are built to take punishment – on your body and on the beach. Mediterranean, Turkey

Fig. 9.6 Losing your pants is embarrassing anywhere: Rusty metal pocket stud points to lengthier saltwater exposure. Mediterranean, Turkey

Apparel   223

Fig. 9.7 T-shirts are standard beach attire and their imprints vital expressions of personality these days. It normally takes a twist of fate (or a spouse without proper understanding) to separate a beloved T-shirt from its owner. Pacific, Mexico

Fig. 9.8 Someone somewhere may be topless. Investigate – you're the beach detective! Mediterranean, Turkey

Figs 9.9 and 9.10 Bikini or bra padding and "push-up" elements can take on a marine debris life of their own. Left, Mediterranean, Turkey; right, Pacific, Hawaii

Fig. 9.11 Topless, OK, but bottomless? Mediterranean, Turkey

Figs 9.12 and 9.13 Beachgoers tend to change their clothes a lot on the beach, and bras and other underwear may simply fail to survive the contortions. Left, Pacific, USA; right, Mediterranean, Turkey

Fig. 9.14 and 9.15 Another explanation for underwear beach litter? Digestive accidents are a staple of exotic beach vacations, and, frankly, few folks are inclined to take soiled underwear back home. Left, Atlantic, USA; right, Mediterranean, Turkey

Fig. 9.16 Why "the law of marine debris aggregation" should apply to underwear is a challenging question. A family that discards together, stays together? Mediterranean, Turkey

Fig. 9.17 With so many shoes on beaches, it's little wonder you'll find a sock or two as well. USA, California

9.2 Footwear

9.2.1 Shoes

Going barefoot in the sand is one of life's great pleasures – at least it was before you read this book. And it's very healthy for your feet. Still, you'll need to wear some kind of shoes going to and from the beach – driving barefoot is illegal in most places, and the path from parking areas to the beach can be strewn with podiatric nightmares. There are a number of reasons for wearing footwear on the beach too. One is marine debris with its assorted sharp-edged, pointed, or yucky objects. In fact, the motto "I've never met a piece of marine debris I'd like to step on" may better portray marine debris and beach litter than any textbook definition. Another reason is sand temperature: it can easily reach 70 °C (160 °F). That's enough to cause burns and have you hip-hopping even without music. Of course, no one should wade in shallow waters, especially on rocky coasts, without tennis shoes or sturdier plastic sandals. Sharp rocks, sea urchins, spiny or toxic snails, and the poisonous spines of certain fish that like to conceal themselves just below the sand surface dictate wearing adequate footwear.

OK, but shoes can't really be a serious marine debris issue, right? Wrong again. Any type of consumer product that is present in the tens of *billions* (simple mathematics again: eight billion people on the planet, each with let's conservatively say 4 pairs of shoes…) will no doubt also end up of the world's beaches. Everyone except the odd marine biologist on the Maldives needs shoes. Shoes wear out, and new models appear. Most people will own a dozen pairs and some – you know who you are – hundreds! Importantly, most shoes today are made of durable plastic, but let's be kinder and more tech-savvy and say "synthetic materials" – which in itself is always a marine debris problem. Of course, beach litter can include the odd leather ("organic") dress shoe, but even this will likely be heavily impregnated with environmentally dubious chemicals and have a rubber heel and synthetic threading.

So how do shoes get to the beach? There are many sources: improperly discarded municipal wastes, beachside dumps, inattentive or blindsided boaters, and, of course, beachgoers themselves. The synthetic material used in shoes is obviously designed to be rugged and is thus slow to decompose. New shoes may float for years and cover thousands of miles on the open ocean (see below). Certain footwear such as neoprene booties for surfers or flippers for divers are especially designed to defy the elements and be saltwater resistant. The durability of sport shoes in the marine environment was demonstrated in

1990, when 21, 40-foot-long containers holding 80,000 top-brand shoes were lost overboard from a container vessel in the North Pacific Ocean [5]. Six months to 1 year later, thousands of shoes washed ashore along a stretch from the Queen Charlotte Islands in Canada to Southern Oregon. Even after drifting thousands of kilometers, many shoes were still perfectly wearable. To find matching shoes, beachcombers held swap meets in Oregon. In 1992, shoes from this mishap were reported to have reached the Big Island of Hawaii. Researchers have used drifting bottles and other items like these shoes to trace ocean currents.

What ultimately happens to shoes that have floated for longer periods and after the softeners and other environmentally unfriendly "additives" have leached out of the plastic? They become encrusted with algae, barnacles and other crustaceans, bryozoans, sea squirts, and many other organisms most non-marine biologists never knew existed. Like other floating objects, shoes can thus play a role in the long-distance dispersal of organisms by "rafting." This includes "alien" species that can cause lots of trouble when floating to faraway seas or landing on remote islands. One scientific paper followed the fate of such shoes and determined that some become so heavily overgrown that they sink to the bottom. There, miles below the surface, in the cold and dark (the average temperature at the deep-sea floor – the largest ecosystem on the planet – is about 3 °C = 37 °F), the encrusting organisms die off, and the shoe rises again to the surface to start the cycle anew.

Can beach sleuths have fun with shoes? You bet! Shoe manufacturing is a deadly serious industry, as one might expect from any billion-dollar industry that services almost every human on the planet. This means strong competition and excruciating efforts to distinguish oneself from the others and even from last year's model (think sports shoes). The ultimate shoe company goal is instant brand-name recognition worldwide. The path is to use selected material compositions, patented gimmick parts, and unique design features. Identifying most brands is child's play because most companies are so proud of their products you can be sure their name and logo are positioned prominently and indelibly. Sizes and other information are also usually stamped into the sole or printed on virtually indestructible tags. They are visible even in footwear long maltreated by the sea. The fetishist will likely be able to tell you the model and year. Finally, the numbers and sizes of encrusting organisms can help biologically interested sleuths estimate the time spent in the water. Organisms that leave behind skeletal parts, such as the white tubes of tube worms or the empty shells of mussels and barnacles, are particularly useful. Enjoy your next shoe find!

Fig. 9.18 Flip-flops and sport shoes: the most common footwear marine debris on beaches. The metal eyelets are the first to go. Mediterranean, Turkey

Fig. 9.19 Outsoles (black) often begin to detach from midsoles. Barnacle encrustation indicates lots of time spent at sea. Pacific, South Korea

Apparel 229

Fig. 9.20 A little bit of foam and plastic can help keep even heavily encrusted (goose barnacles, blue mussels) shoes afloat for ages and across whole oceans.

Fig. 9.21 Upper and lower shoe have parted ways. The secret to "air" soles (namely, holes!) revealed. Mediterranean, Turkey

Figs 9.22 and 9.23 Plastic soles and toe caps are always hardier than the fabric upper parts. Mediterranean, Turkey

Figs 9.24 and 9.25 Sometimes a logo, special design feature, or sturdy eyelets help identify shoe remains. Detached uppers can take on a marine debris life of their own thanks to multiple layers and fancy stitching. Mediterranean, Turkey

Fig. 9.26 and Fig. 9.27 Tough soles mean long-lived marine debris. Company logos are a great help to beach detectives. Mediterranean, Turkey.
At least I think this is part of a trick shoe sole (run faster, jump higher, etc.?). Caribbean, Cuba

Fig. 9.28 Rubber heals, nails, synthetic stitching, colors, and treatment with chemicals mean that leather shoes are not really "organic." Pacific, Mexico

Apparel 231

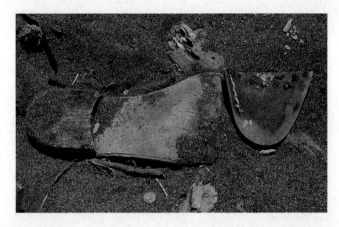

Fig. 9.29 Rusty spots show that dressier shoes are both stitched and nailed together. Mediterranean, Turkey

Fig. 9.30 Remnant stitching and low-tech "air" soles – the final stage of most shoes. Mediterranean, Turkey

Fig. 9.31 "Heel," said the beachcomber to no dog in particular. Mediterranean, Turkey

Fig. 9.32 Small children (and their harried parents) are the only ones who might be excused for losing shoes on the beach. Atlantic, USA

Fig. 9.33 Alas, this ballerina shoe shall pirouette no more. Adriatic, Italy

Fig. 9.34 Footwear specifically tailored for water sports is booming and makes for especially rugged marine debris items. Pacific, Mexico

Fig. 9.35 Rubber boots belong to the working outfit of many marine jobs. Mediterranean, Turkey

Fig. 9.36 Encrusted with tubeworms and oysters, the final decomposition stages of a rubber boot. Rather icky as marine debris specialists would say. Pacific, Japan

Fig. 9.37 A wooden shoemaker's form – one of the more unusual footwear-related marine debris items. Red Sea, Jordan

9.2.2 Flip-Flops

Flip-flops are such iconic and ubiquitous beachwear that they earn their own little subchapter. They apparently got their name from the smacking sound they make when their soles alternately hit your heels when walking. That's onomatopoetic for those of you who did well in English class. While they do offer city folk a precious chance to air out their feet, they are inherently flimsy (but, looking at some name brands, not necessarily cheap). That also helps explain why you find so many on beaches. The "drag" of dry sand and the suction of wet sand are more than most flip-flops, thongs, and the like can handle. Their "weak links" soon give out. Who wants to take a dismembered flip-flop back home – and there is little use in saving the orphaned, intact member of the pair, right? Why are flip-flops a marine debris issue? Beyond their numbers, they float very well indeed (think rafting of "alien" species) and usually contain abundant additives (softeners) that ultimately leach out into the environment. Interestingly, even old flip-flops can be upscaled: enter "flip flop mat" into your computer's search engine! Another company in Kenya recycles 400,000 of them a year into toys, functional items, and art, employing 50 artists and supporting hundreds of people [6]. If you are thinking about wearing flip-flops during a beach cleanup, reconsider: the typically worn and thinly compressed heel material (and open-toe design) may not do a good job of protecting you from being injured if you step on glass, not to mention syringes, sea urchin spines, etc.

Fig. 9.38 Flip-flops and plastic sandals in every size, color, and style. Iconic, ubiquitous, and unisex, sold along every tourist beach in the world. Mediterranean, Turkey

Fig. 9.39 Flip-flops are typically abandoned on site when one of the thong bands detaches. Mediterranean, Turkey

Fig. 9.40 They always seem to break at the same places. The flip-flop on the top is in good marine debris company. Pacific, Mexico; Caribbean, Cuba

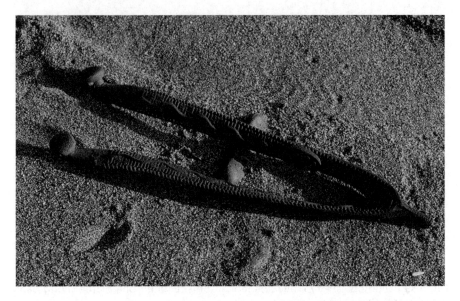

Fig. 9.41 The suction force of wet sand and the "drag" of dry sand are more than most thongs can hande (peg on right missing). Pacific, Mexico

Fig. 9.42 and Fig. 9.43 The heels of cheap foam soles quickly wear through (left). The two adjoining rear thong pegs are still attached; toe peg is lost. Caribbean, Cuba.
The sturdy peg reveals this to be a flip-flop marine debris remnant. Caribbean, Cuba

Fig. 9.44 Flip-flops: in some countries valuable enough to be repeatedly repaired. Pacific, Mexico

Fig. 9.45 Every flip-flop ultimately meets its match in the beach environment. Mediterranean, Italy

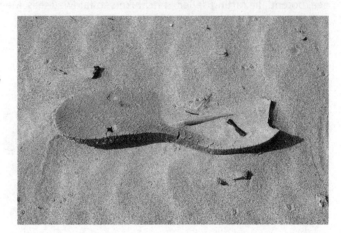

Fig. 9.46 and 9.47 The characteristic recesses or holes for the thong pegs help beach detectives recognize even small flip-flop remains. Mediterranean, Turkey; Caribbean, Cuba

Fig. 9.48 Encrusting tubeworms: this undamaged sandal has spent time floating at sea, potentially rafting "alien" hitchhikers to faraway seas. Mediterranean, Turkey

Fig. 9.49 Goose barnacles (arrows) and the size of other "fouling" organisms can help estimate the age of some soles. Pacific, Hawaii

Fig. 9.50 No, not an alien. The thong pegs are clearly more robust than the weathered soles. Mediterranean, Turkey

Apparel 239

Fig. 9.51 You'll have to look twice to recognize some flip-flop soles (see peg remnant on top). Mediterranean, Turkey

Fig. 9.52 Paired flip-flops usually mean a nearby owner. Batman has abandoned both of these because one is damaged (arrow). Be shabby chic: collect the good ones to create funky mismatched pairs. Mediterranean, Turkey

Fig. 9.53 No, for once this is not "the law of marine debris aggregation." Paired, intact shoes along the waterfront usually mean owners swimming nearby (and probably watching you warily). Mediterranean, Turkey

Fig. 9.54 Collective sandal suicide? Some answers elude even the most experienced beach detective. Pacific, South Korea, Jeju Island

Fig. 9.55 Even old flip-flops can be upscaled, for example, into colorful doormats. California

9.3 Gloves

What do you need to know about gloves? Basically one thing: if someone used them, it's because they handled something they didn't want to touch with their bare hands – often for good reason. For most of us, it may simply be to avoid shriveled, "pruney" fingers while dishwashing, but for others, it's handling toxic chemicals, contaminated materials, vile wastes, sharp objects, or other unsavory items and fluids. And there are gloves for every such disagreeable purpose. This means an incredible variety of work and safety gloves as well as industry gloves. Some need to be abrasion-resistant and others chemical-resistant, cut-resistant, vibration-dampening, and heat- or impact-resistant. Not to mention insulated rubber gloves for electrical protection with various classes, colors, and maximum use voltages. Depending on the duty, the glove materials can be cloth, leather, latex, rubber, nitrile, PVC, PVA, neoprene, Kevlar, butyl/Viton-butyl, and other imaginative materials as well as every imaginable combination thereof. Other features beachcombers might look for are gloves with fingers versus mitts, gloves with/without gauntlets or safety cuffs, and disposable versus non-disposable gloves. All this means gloves come in a truly astounding variety, with two of my favorites being "frisk duty premium goat-grain cut resistant law enforcement gloves" [7] or "featherweight anti-static filament nylon and touchscreen cleanroom gloves."

Glove labels on "serious" gloves sometimes provide codes and other information related to internationally recognized standards. Category 1 gloves, for example, are of simple design for minimum risk, category 2 for intermediate risk, and category 3 of complex design to avoid irreversible injuries or mortal risks. Keep an eye out for multilayered or so-called palm-coated gloves, whose added protection can only mean nasty duty. Gloves designed to protect you against chemicals are a case in point. And you need the right glove for the right chemical. Such gloves can be labeled, in the European Union, for example, with a pictogram and a three-digit code specifying three standard chemicals. Depending on glove quality, they can shield your hands for between 10 min ("class 1") and 8 h ("class 6") before you and your hands are in real trouble. This implies that gloves can take some beating, but all eventually have to be discarded and replaced. For more information on codes, pictograms, and the chemicals that various materials can ward off, check the Internet [8]. Needless to say, if a beached glove has a pictogram related to protection against microorganisms (biohazard; Fig. 1.12), you can rightfully be worried.

But that's more than you'll probably need to know about gloves as a marine debris issue. Perhaps a good rule of thumb (pun intended) is the heftier the glove, the more industrial the task – and the more wary you should be upon finding one on the beach. At any rate, gloves are an item *not* left behind by beachgoers. The question to ask is how many gloves are you finding and what might that be telling you about a nearby industry somewhere offshore, along the coast, or upstream on the nearest river. And if that industry is "losing" gloves, then what else might it be discarding that's ending up on your beach? With some luck, you'll encounter a "logo" glove imprinted with a company logo and address or customized promotional message. This means you can send a nice message back in return or your beach cleanup coordinator can "hand" the matter over the authorities.

Fig. 9.56 Heavy-duty palm-coated glove. Wear and tear indicate function far beyond mere dishwashing. Mediterranean, Italy

Fig. 9.57 Every extra plastic coating, like on this lovely two-tone model, indicates dabbling in unsavory liquids. Pacific, South Korea

Apparel 243

Fig. 9.58 Overgrowth by bivalves and tubeworms: this simple glove has washed in from the sea. Mediterranean, Italy

Fig. 9.59 Disposable glove, the kind most often used in research labs, or by your proctologist. Use gloves to collect gloves during beach cleanups. If you find one, keep an eye out for more: they could mean improperly disposed of medical wastes. Mediterranean, Turkey

Fig. 9.60 The "ribbed" fingertip provides the final glove clue for experienced beachcombers. Mediterranean, Italy

9.4 Hats and Caps

Among the three things you need most on the beach – sunscreen, water, and hat – hats may be the most neglected. In cold weather, we lose the most heat through our heads. In summer, excessive heat tends to, well, fry our brains. Not to mention burn our noses and for some, the old bald spot. While you can usually borrow sunscreen or mooch a drink, who will lend you a hat? After all, a hat is often the key beach attire fashion statement, something you don't lend to just anyone. For some, it may also be the largest clothing item they have on.

What kind of headwear is the best for beach duty? The bigger, the better! Forget your headband, bandana, or mere visor. A sombrero is the perfect ticket against the sun but can be a real liability on windy days. Something along the lines of an "Aussie" tennis-like hat seems just about perfect: white color (reflects sunlight), a brim that goes all the way around (protects your nose, ears, and the back of your neck!), a green underbrim (reduces glare), a few small holes or patch of mesh on top for air exchange, a bit of adjustability if the wind picks up, and no metal eyelets or zippers (saltwater means ugly rust spots).

That said, what do folks wear most on the beach: baseball caps. The technical expression for this solution: "suboptimal," regardless whether the bill is worn facing forward or, please spare me, backward. Of course, that makes baseball caps the most common hat debris, even in countries where baseball as a sport is largely unknown. Depending on where you vacation, however, you may also find various other types of headwear, for example, "personal protective equipment" or – for us simple folks – hard hats. Classical locations for such debris are beaches with offshore industries such as oil rigs. Hard hats are built to protect heads against sharp or blunt blows, which means they are robust indeed and make for long-lived marine debris. You'll recognize them by the hard outer shell and the often adjustable internal suspension system that helps absorb impacts. If the surf hasn't maltreated them too much, they might still bear chinstraps, sweatbands, and the like. Some more high-tech types are even equipped with visors, headsets, or lights. And luckily for beach detectives, hard hats are often company equipment and bear a company logo or name. That makes it easy to complain to the right people!

Fig. 9.61 OK, beach detective, what's this? Yep, the final marine debris stage of a baseball cap: the internal plastic part of the bill. Shape, stitching holes, and traces of stitching are the clues. Mediterranean, Turkey

Fig. 9.62 Nice find (if you like being a walking advertisement) – hats as beach litter usually mean lots of wind. Perfectly intact apparel counts among the "Big 10" desirable marine debris items. Mediterranean, Turkey

Fig. 9.63 If turquoise is your favorite color, this cap may be only one good washing away from being yours. Mediterranean, Turkey

Fig. 9.64 Decomposing but still advertising. The cloth has begun to detach from the bill – maybe making it an even more hip, shabby-chic find? Mediterranean, Turkey

Fig. 9.65 All the parts are still there, but the cloth-less bill is nearly separated from the cap, and the plastic adjustment band is broken. Mediterranean, Turkey

Fig. 9.66 Hard hat: someone may be doing dangerous and dirty work off "your" beach. What else might that industry be losing? Encrusting organisms mean time spent at sea. Built to take major abuse and durable as marine debris. Pacific, South Korea

Fig. 9.67 Hard hat: advanced stage of decomposition. Outer shell gone, inner fitting (suspension) and adjustment mechanisms are tough and long-lived. Pacific, South Korea

Fig. 9.68 Lightweight, airy, mostly organic. A cool find in any condition. Mediterranean, Turkey

Fig. 9.69 Cheap shower caps are typically provided by every hotel for folks who don't like to get their hair wet. Mediterranean, Slovenia

References

1. Dockterman E (2012) EcoChic: How U.S. clothing brands are getting greener. Time. 20 August 2012
2. The Sustainable Apparel Coalition (SAC) https://apparelcoalition.org/
3. Mediterranean Association to Save the Sea Turtles (Medasset) http://fromwastetowear.medasset.org/en/socks/
4. Dris R, Gasperi J et al (2016) Synthetic fibers in atmospheric fallout: a source of microplastics in the environment? Mar Pollut Bull 104(1–2):290–293 https://doi.org/10.1016/j.marpolbul.2016.01.006
5. Ebbesmeyer CC, Ingraham WJ Jr (1992) Shoe spoil in the North Pacific. Earth Space Sci News (EOS) 73(34):361–365
6. Ocean sole: flip the flop http://oceansole.co.ke/
7. Superiorglove http://www.sglive.pairsite.com/frisk-duty-premium-goat-grain-cut-resistant-law-enforcement-gloves
8. http://www.crudehandsgloves.com/wp-content/uploads/2016/03/EN-guide_EN.pdf

10

Water Sports

Why do we go to the beach and what do we do once there? Lying in the sun all day is downright deadly says the dermatologist. Reading, eating, and drinking are all very fine but can't really fill up the whole day, much less a longer vacation. Things can get boring fast. This means you'll need some action, and this means sports. You can work up a sweat directly on the beach: kicking a soccer ball around or throwing a football, setting up an impromptu beach volleyball court, trying simple racquet games, or throwing a "Frisbee." A major drawing point of heading to the shore, however, is actually getting into the water. Most of the exercise people get on a beach vacation is therefore water-bound.

What does this all have to do with beach litter or marine debris? In our high-tech world, we seldom go into the water wearing only our swimsuits. The water sports industry is a multi-billion dollar endeavor that offers a truly endless array of gear and equipment that enables us to swim faster, dive deeper, or skim the surface at breakneck speed. The relevant fact in our beach litter context is that most folks come back out of the water with fewer items than they originally waded in with. Sometimes an honest wave does the trick even if you went in wearing only a swimsuit. Tip: pull those strings tightly and never turn your back to the ocean! Snorkelers and divers also know all too well that spending time in the water makes you cold and stupid fast. And in this state, you tend to make benumbed decisions and lose things quickly.

Water sports equipment is built to take punishment and to withstand saltwater, strong sunlight, water pressure, and wave action. This means that, next to fishing gear, sports equipment makes for very long-lived marine debris. Nature has two strategies to help organisms withstand the brute forces of the sea. They can be either very hard (think barnacles or limpet shells) or be very flexible (think kelp or other algae). You'll see the same strategies in sports equipment. Surfboards and other equipment designed to support the weight of a frantically active human body and keep us out of the water are typically very hard. Other items such as swimming fins are very flexible. Each takes optimal advantage of the properties of water to help us best achieve our sporty ambitions.

Finally, most water sports equipment, such as surfboards, float very well indeed. If you lose your snorkel and it rises to the surface, you might be lucky and find it. If not, ocean currents might transport it halfway around the world for someone else to find. If your equipment is neutrally balanced in the water (think underwater camera), it can still easily drift off and be lost. Sports equipment heavier than water can seldom be retrieved, especially considering the often poor visibility and depth limits of snorkeling and diving (let's say 30 m) compared to the average depth of the oceans (4000 m). Importantly, items that sink to the bottom tend to decompose even more slowly: colder temperatures, less water movement, fewer bacteria, and little or no sunlight considerably retard the deterioration of any material.

If you are out scouting the beach for litter, you might actually be thrilled at finding intact sports equipment, which is among the "Big 10" desirable or valuable marine debris items (glass buoys, message-in-a-bottle, stranded watercraft, money, watches, jewelry and other valuables, dry drug packages). The fact that it's made to be durable means it can long remain in good shape. If it's relatively new, undamaged, and not overgrown by encrusting critters, chances are good you can use lost sports equipment to embark on your own next adventure: what a great way to get free stuff and reduce marine debris at the same time!

Fig. 10.1 Most folks come out of the water with less than they went in with, and divers and snorkelers tend to lose much more than they find, including essential gear. Mediterranean, Turkey

Fig. 10.2 The soft rubber parts of swimming fins are their Achilles heel. Most folks tend to discard the remaining good fin too. Mediterranean, Turkey

Fig. 10.3 The toughest parts of flippers (boot and reinforced edges) survive the longest. Red Sea, Jordan

Fig. 10.4 With some luck you'll find a complete and undamaged mask and snorkel set. Mediterranean, Turkey

Fig. 10.5 In most cases, however, you'll find them separately, damaged or simply not appetizing enough to stick in your mouth. Mediterranean, Italy

Fig. 10.6 and Fig. 10.7 The soft "rubbery" snorkel mouthpieces can take the most punishment and stick around the longest. You'll find them in every flexible shape and form. But even the sturdiest snorkel mouthpieces eventually decompose. Mediterranean, Greece

Fig. 10.8 One can only hope the damage occurred on the beach rather than on someone's face. Mediterranean, Turkey

Fig. 10.9 The silicone "skirts" and nose pockets of good masks are very pliable. Glasses tend to pop out and plastic frames and headstraps to detach. Mediterranean, Greece

Fig. 10.10 Missing pieces are bound to be living their own marine debris life somewhere close by. Mediterranean, Greece

Figs. 10.11 and 10.12 The most common underwater sports items found on beaches are the variously designed pieces that hold the mask and snorkel together. Mediterranean, Turkey; Pacific, Hawaii

Fig. 10.13 and Fig. 10.14 Beachcombers and sometimes beach detectives have to look twice to identify sports equipment. Robust adjustable strap fragments typically belong to diving fins. Pacific, Hawaii. Flimsier straps tend to come from diving masks. Mediterranean, Greece

Fig. 10.15 Freshly purchased water sports equipment is often unpacked on site, making its packaging a common type of marine debris. Leave it at the store for proper disposal. "Safe nature?" This packaging wasn't designed in an English-speaking country. Mediterranean, Turkey

Fig. 10.16 Swimming goggles, equally at home in swimming pools and the sea. The clouded lenses indicate abrasion in the surf zone. Mediterranean, Turkey

Fig. 10.17 and Fig. 10.18 Single eyepiece and adjustable nose strap of swimming goggles. Stepping on marine debris, especially on gravelly beaches, hastens the step to microplastic. Mediterranean, Turkey. Either of these items alone might have been difficult to identify as a goggle eyepiece and its foam padding. Strap remnants are a good clue. Mediterranean, Turkey

Fig. 10.19 This swimming goggle packaging contains more plastic than the product itself. Mediterranean, Turkey

Fig. 10.20 A gust of wind or misplaced kick can quickly turn a beloved family beach ball into a piece of marine debris. Mediterranean, Turkey.

Fig. 10.21 A product designer's dream combining family fun and advertising. Mediterranean, Turkey

Fig. 10.22 Cheap, lightweight balls mean thin and flimsy: a short-lived useful life but long marine debris career. Mediterranean, Turkey

Fig. 10.23 Even sturdy soccer balls eventually succumb to the unforgiving beach environment. Mediterranean, Turkey

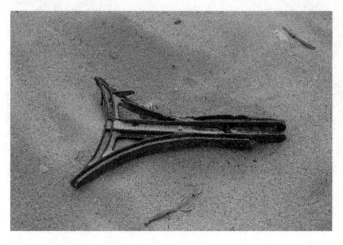

Fig. 10.24 The racquets for ubiquitous beach ball games are also typically lightweight with predefined weak spots. Caribbean, Guadeloupe

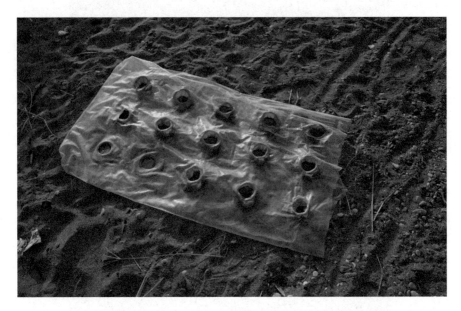

Fig. 10.25 Water and vacation mean air mattresses. No matter how expensive or sophisticated, a mere pinprick hole or sudden gust of wind away from becoming marine debris. Mediterranean, Turkey

Fig. 10.26 Leaving behind colorful, shredded but easily identifiable remnants. Mediterranean, Turkey

Fig. 10.27 and 10.28 "Personal flotation devices" – often a euphemism for cheap, unsafe, and short-lived swimming aids, usually for kids (despite the fine print warnings). Mediterranean, Turkey. Parents buy them; children lose them. They come with extensive instructions and warnings, none of which mention not leaving them on the beach as future marine debris. Mediterranean, Turkey

Fig. 10.29 and Fig. 10.30 Swim noodles, among the most ubiquitous of modern "swim aids" for young and old. Also known as pool noodles or woggles. The outer layer of this broken noodle has already bleached in the sun. Mediterranean, Turkey. Made of plastic (polyethylene foam), swim noodles eventually break apart into bleached, ever-smaller fragments. They can often be identified by their color and rounded parts (and by other fragments nearby: "law of marine debris aggregation"). Mediterranean, Turkey

Fig. 10.31 An intact surfboard of any kind is something you'll actually be happy to find. It's on the beachcombers' list of the "Big 10" desirable marine debris items. With luck, this means a short career as beach litter. Mediterranean, Turkey

Fig. 10.32 and Fig. 10.33 Kudos to someone for trying to prolong the lifespan of this board ("repair," as one of the 6 "R"s); thumbs down for ultimately abandoning it as marine debris. Mediterranean, Turkey. Ouch! It takes a lot to break a surfboard but, assuming the surfer survived, also a lot of chutzpah to simply leave it on this boulder beach. Pacific, California

Fig. 10.34 Surfboard tip or nose guard – designed to protect surfers and their boards – is probably the smallest surf-related item you'll find. Pacific, Hawaii

Fig. 10.35 Cheaper "styrofoam" surf- or bodyboards break more easily and are more common marine debris items than their fiberglass or epoxy cousins. Mediterranean, Turkey

Figs. 10.36 and 10.37 When rafting or kayaking, what breaks most often and becomes marine debris? Cheap paddles! Mediterranean, Turkey

Fig. 10.38 Congratulations, you've bagged another one of the "Big 10" desirable marine debris items. If no one is around far and wide, it's not tethered or stored neatly upside-down, you should at least get a finder's fee out of it. If it's damaged and water-filled and very ugly to boot, well, then it's just marine debris (see also Chaps. 3 and 12 (metal and wood) for larger vessels). Mediterranean, Turkey

11

Fishing Gear

Recorded history shows that fishing has always been and continues to be our biggest activity in the world's oceans: the main thing we want from the sea is its fish. Seafood is very healthy we are told (and the main source of protein for many folks), but the fishing industry (and we as consumers) pay a high ecological and societal price for this obsession with fish and fishing. Overfishing threatens the populations of most commercial species, with an estimated 70% being fully or overexploited, yet per capita fish consumption is projected to further increase on all continents [1]. This has led to feverish, fire brigade-like, rearguard measures: fishery closures, ever-tighter fishing seasons, size limits, stricter licensing, stronger penalties, a reassessment of statistical analyses, and the precautionary establishment of marine protected areas (always one step behind, one step too late) [2, 3]. The bycatch problem adds insult to injury: species that we do not eat or even want to catch are "accidentally" captured along with the "target" species. Such bycatch and discards – an estimated annual average of 27 *million* tons or about one-quarter of all the fish we catch – end up thrown back dead into the sea [4]. High costs also include the immense subsidies given to fishers to offset the fuel expenses for reaching ever more remote fishing grounds. Politically, explosive confrontations involving fishing grounds and "territorial" waters threaten peace from the Adriatic Sea in the Mediterranean to the Atlantic ("cod wars") to the South China Sea in the Pacific [5, 6].

In the present context, we are more interested in the downright shocking marine debris dimension to fishing. Most human activities and industries seem to contribute substantially to marine debris and beach litter. Fisheries are not only no exception but often the worst offenders. Why? Firstly, the oceans are brutal, and fishers lose lots of equipment. Secondly, the oceans are a practical on-the-spot place to get rid of lots of waste. Thirdly, lost or abandoned gear is especially evil because of its amounts and because of its properties: durable and made to kill.

For millennia, humans have devised equipment to variously trick and ever more efficiently capture fish, squid, crabs, lobsters, and clams. Not to mention whales, dolphins, and sea turtles. Entire multilingual dictionaries have been written to encompass the incredible array of fishing gear in use today [7]. This ranges from single hooks to football field-sized trawl nets and 100-km-long longlines and driftnets ("walls of death"). Combined, these are let into the water in untold millions.

As you can guess, not everything slipped into the water comes back on board. Hooks snag, lines snap, nets tear, buoys detach, traps, pots, and fish crates and tackle are washed overboard, and indeed entire fishing vessels with all their gear are sometimes unfortunately lost at sea. An estimated 650,000 tons of fishing gear are left in the oceans each year. In the Northeast Atlantic alone, about 25,000 nets (total length 1250 km) are lost or discarded annually [8]. And let's not forget the millions of kilometers (that's millions of miles too!) of monofilament fishing lines lost by anglers each year. Fisheries are therefore a major source of marine debris, and various gears such as fishing line, lures, and nets are separately listed in most beach cleanup data cards [9]. On some remote islands, the commercial fishing equipment that has washed ashore has apparently even made landing on beaches with small boats difficult. Such gear includes enormous nets weighing up to several tons – beyond the scope of any volunteer beach cleanups.

The ocean is a cruel environment, and the things we put into it on purpose are typically made to withstand brutal winds, surging waves, blinding sunlight and/or icy waters, corrosive saltwater, merciless winches, and the weight of encrusting marine organisms. This is especially true of fishing equipment, which beyond defying the above forces is designed to reel in frantically struggling, powerful fish or heave entire multi-ton schools of fish out of the water. This calls for stainless steel, impact-resistant plastics, high-tech synthetic ropes and lines, and numerous other robust components in endless combinations. This means two things: fishing gear is expensive and valuable, and fisheries-related marine debris is highly resistant to degradation.

What are the threats? Beyond being an eyesore, lost fishing gear poses a major threat to marine life. This menace is fourfold: continued capture of target species, entanglement of and ingestion by nontarget species, and the destruction of seafloor communities (smothering). Lost nets ("ghost nets") and traps can continue to kill the fish and invertebrates that they were designed to capture for years after having become separated from their owners' buoys, lines, or GPS coordinates. Entangled wildlife includes everything from snagged birds, drowned or amputated turtles, strangled sea lions, and horribly mutilated cetaceans (whales and dolphins). A significant proportion of certain whales such as humpbacks and right whales today are scarred by fishing gear: the many severely entangled animals, doomed to a slow and painful death, have prompted a special disentanglement program by the International Whaling Commission [10]. As far as ingestion goes, from an evolutionary perspective, everything floating on or drifting in the sea was organic and worth taking a nibble at or swallowing whole. Sea life (and most life come to think of it) is simply designed to inspect and try to consume almost everything it encounters – including fisheries-related items. Finally, derelict fishing gear can sink to the bottom and smother life on the seafloor. In coral reefs, such gear also snags the coral colonies and tears them apart when storms hit the coasts.

Fishery-related marine debris poses a human health hazard as well. Entangling your propeller in an abandoned fishing net or lost line can leave you adrift and unmaneuverable on the high seas. And probably just at the worst possible moment, like approaching nightfall or a typhoon on the horizon (or both, Murphy's Law). Underwater, most fishing lines are practically invisible: this poses a serious risk to snorkelers and divers. That – and not to ward off inquisitive sharks – is the main reason why every diver should always carry a sharp knife. On the beach, fish hooks are just as bound to snag your foot as they would a fish in the water. Finally, removing heavy and partially buried nets and ropes from the beach can be a major chore: watch your back. And wear gloves: most fishery items have spent lengthy periods at sea before drifting ashore and are covered with very sharp-edged barnacles, mussels, and other encrusting species.

What can experienced beachcombers contribute to the cause? For one, most fishery items are expensive. This means that traps and nets – or the buoys to which they are attached or that mark deployment sites – are often individually marked. This can range from owner or ship names to color-coded buoys and flags. All are designed to help fishers relocate and distinguish their gear from that of adjoining competitors. It can also help identify lost gear. For example, the fishing gear entangling whales swimming off Hawaii can sometimes be traced back to nets or traps originating off the North Pacific coast of the USA or Canada. Such information can be used to identify problematic gear types, localize entanglement hotspots, and help redesign equipment or develop new deployment strategies. So don't forget to at least take a photo.

What are the alternatives to harmful fishing gear and the debris they become? Destructive fishing practices can be restricted (dredging). Gear that indiscriminately and "unintentionally" captures wildlife can be banned or their modes of deployment changed (shorter and deeper longlines). Equipment can be redesigned (predetermined breaking points or failure joints, with marine debris and wildlife in mind). Importantly, efforts should be supported to have derelict gear collected at sea by fishermen and recreational boaters. Reward payments by the pound/kilogram would be mighty motivating! A Global Ghost Gear Initiative has been started [11], an International Smart Gear Competition created [12], and a Marine Stewardship Council (MSC) certification scheme implemented [13]. On a more local level, fishing line recycling stations are popping up at marinas and tackle shops everywhere. The recycled products made from them range from park benches to fish habitats. Over a recent 20-year period, one company alone recycled more than 9 *million* miles of fishing line [14]. By the way, there are successful businesses regenerating the nylon in recovered fishing nets to make new "sustainable" textiles for use in clothing such as socks under the motto "from waste to wear" [15].

What can you and I do? How about buying a pair or two of those socks and telling our friends? The anglers among us can replace our stainless steel or alloy hooks, which forever remain embedded in escaped fish and snagged turtles, etc., with standard hooks that will eventually rust away. We can replace our brittle lines before they tear at sea, and we can properly discard them at the nearest fishing line recycling station (check your marina, local fishing pier, or tackle and bait shop). You can also specifically ask for biodegradable fishing line at your next purchase. Finally, be sure to collect and correctly dispose of any fishing hooks and lines you find on the beach, even if you are not participating in an official beach cleanup.

Fig. 11.1 Fishing is one of the last "hunter-and-gatherer" domains in modern life, and fish hooks, swivels, lines, and other fishing paraphernalia abound on beaches. Pacific, Hawaii

Fig. 11.2 The "law of marine debris aggregation" means you'll often find (or step on) more than a single hook. Mediterranean, Turkey

Fig. 11.3 This lost triple hook may still be freshly packaged, but that won't spare the hapless barefoot vacationer. Atlantic, USA

Fig. 11.4 Lost hooks can get ugly fast. That's actually good for snagged and escaped fish (and sea turtles, etc.): the hooks corrode away. Anglers should consider replacing their stainless steel hooks with such standard products. Mediterranean, Slovenia

Figs. 11.5 and 11.6 Even experienced beachcombers will sometimes have to look twice or turn an item over to see that it is actually hook-studded. Mediterranean, Turkey

Fig. 11.7 Lures designed to catch fish or slippery squid are guaranteed to latch onto your foot equally well. Although hooks typically rust away first, pick any lures up gingerly during beach cleanups. Mediterranean, Italy

Fig. 11.8 Beware of artificial-looking, wriggly wormlike things on the beach. They may fool fish, but they shouldn't fool you: such fishing lures have hooks hidden somewhere. Atlantic, USA

Fig. 11.9 and Fig. 11.10 In most parts of the word, fishing lines are not carefully wound on expensive reels. Such handheld spools are often lost along with hundreds of meters of line. One estimate of how long fishing line remains in the environment: 600 years [16]. Mediterranean, Turkey.
Millions of miles of fishing line are lost or cut every year. You'll find them as tangles on the beach or, worse, wrapped around the heads and feet of drowned birds or cutting through the flippers of sea turtles. Nearly invisible underwater, they pose a threat to snorkelers and divers too. Mediterranean, Turkey

Fig. 11.11 Small lead sinkers can be ingested by wildlife. So much so that lead pellets for shotgun shells have been banned in many places after waterbirds showed lead poisoning symptoms (not from being shot but from eating them). Mediterranean, Turkey

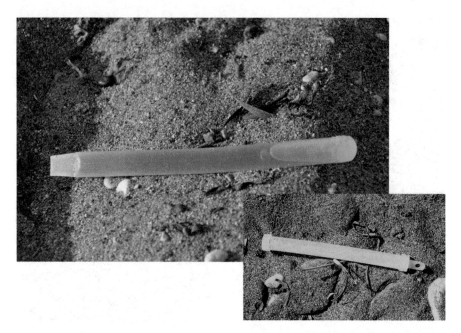

Fig. 11.12 and Fig. 11.13 Pen-sized, fluid-filled plastic objects are usually "glow sticks" to attract fish at night. Bending them snaps a capsule floating inside, and the mixture emits light for several hours. Glass, plastic, and chemical components – all designed for onetime use and commonly discarded at sea. Mediterranean, Turkey.
You'll also recognize them by their top ends, which are variously designed to attach to fishing lines. Mediterranean, Turkey

Fig. 11.14 And each stick comes in wrapping which is also more often than not discarded overboard. "Bend, snap, and shake": more than enough clues for any beach detective! "Shake, rattle, and roll" produces less marine debris. Atlantic, England

Fig. 11.15 and Fig. 11.16 Ropes are a fundamental aspect of life at sea and made to take extreme punishment. This thick mooring line nonetheless met its match and became marine debris (caliper in center open to 1 cm). Mediterranean, Slovenia. Barnacle-covered close-up shows the frayed, intricately interwoven structure of most ropes. Mediterranean, Slovenia

Fig. 11.17 From heavy-duty rope to microplastics (microfibers): each strand itself consists of further strands that ultimately unravel, become brittle, and break. Mediterranean, Slovenia

Fig. 11.18 Commercial fishing gear, either lost or discarded at sea, can weigh tons and poses a severe entanglement threat to even the largest marine organisms, whales. During beach cleanups, such marine debris requires heavy machinery to remove. Pacific, South Korea

Fig. 11.19 Like icebergs, the bulk of commercial fishing gear may be buried in the sand. The more you pull (watch your back!), the longer and heavier it gets. This is nothing for inexperienced beach cleanup participants. Atlantic, Scotland

Fig. 11.20 Commercial fishing gear on beaches is typically a tangle of netting, ropes, floats and weights – all designed to withstand the strongest forces on the planet and therefore long-lived marine debris. Pacific, South Korea

Fig. 11.21 and Fig. 11.22 The hundreds of small buoys used to keep fishing nets upright in the water become marine debris when the lines eventually cut through them. Mediterranean, Slovenia.
Chances are good that any plastic, bagel-like fragment is a fishing buoy remnant. Mediterranean, Turkey

Figs. 11.23 and 11.24 Buoys come in many shapes, sizes and attachment possibilities, eventually becoming abraded, aged and barely recognizable after rolling in the surf zone. Identify the old ones based on the new ones no doubt lying around close by. Pacific, Hawaii

Fig. 11.25 "The law of marine debris aggregation" means you'll often find large accumulations of most items, including (apparently intact) fishing floats. Mediterranean, Greece

Fig. 11.26 And then there are the whoppers required to keep massive nets spread out and afloat. The bottom side of this one is heavily encrusted with barnacles, indicating the original orientation in the water. Pacific, South Korea

Fig. 11.27 Rough seas and brutal winches eventually crack even the largest and sturdiest buoys. Pacific, South Korea

Fig. 11.28 Often it's the lines that tear first, with the buoy being damaged when pounded against rocky shores. The reinforced "eyes" on fishing and mooring buoys tend to survive the longest. Mediterranean, Turkey

Fig. 11.29 These plastic "fish throats" or funnels from fish traps or eel pots are one-way entrances for the hapless prey. Pacific, South Korea

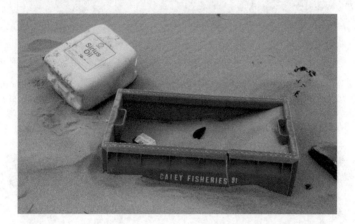

Fig. 11.30 Fish boxes. Rugged, made to be thrown about and stacked ahigh. Some are clearly marked to provide beach detectives with vital clues. Atlantic, Scotland

Fig. 11.31 Other fish boxes have giveaway features for beach sleuths: perforations along the bottom sides let excess water escape. Mediterranean, Italy

Fig. 11.32 Check out your local harbor to see where the many fish boxes come from. Catching tons of fish means heaving them about in manageable units, such as this array atop a fishing vessel. See Chaps. 5 and 12 (foamed plastic and wood) for other varieties. Mediterranean, Slovenia

Fig. 11.33 Aquaculture (mariculture) is touted as solving the "proteins from the sea" problem by relieving overexploited fisheries but is also a major source of marine debris. This tangle of plastic washed ashore from a mussel farm. The mussels grow in such long plastic "socks" suspended in the water. Mediterranean, Italy

References

1. FAO (2016) The state of world fisheries and aquaculture 2016. Contributing to food security and nutrition for all. Rome. 200 pp. www.fao.org/3/a-i5555e.pdf
2. Longhurst A (2010) Mismanagement of marine fisheries. Cambridge University Press, Cambridge 320 pp
3. Finley C (2011) All the fish in the sea. Maximum sustainable yield and the failure of fisheries management. University of Chicago Press, Chicago 224 pp
4. Alverson DL, Freeberg MH et al (1994) A global assessment of fisheries bycatch and discards. FAO fisheries technical paper no. 339. Rome, FAO, 233 pp. http://www.fao.org/docrep/003/T4890E/T4890E00.HTM
5. Fravel MT (2011) China's strategy in the South China Sea. Contemp SE Asia: J Int Strateg Aff 33(3):292–319
6. Xie Y, Luo S (2015) Fishing for ways to de-escalate South China Sea tensions. The Diplomat. https://thediplomat.com/2015/08/fishing-for-ways-to-de-escalate-south-china-sea-tensions/
7. Commission of the European Communities (1992) Multilingual dictionary of fishing gear, 2nd edn. Fishing News Books, Oxford 333 pp
8. Animal World Protection. https://www.worldanimalprotection.org/our-work/animals-wild/sea-change
9. Ocean Conservancy. http://act.oceanconservancy.org/site/DocServer/ICC_Eng_DataCardFINAL.pdf?docID=4221
10. International Whaling Commission. https://iwc.int/entanglement
11. Global Ghost Gear Initiative. https://www.ghostgear.org
12. https://www.worldwildlife.org/initiatives/international-smart-gear-competition
13. Marine Stewardship Council. https://www.msc.org/
14. Ocean Conservancy (2011) Tracking trash: 25 years of action for the ocean. https://issuu.com/oceanconservancy/docs/marine_debris_2011_report_oc
15. Mediterranean Association to Save the Sea Turtles (Medasset). http://fromwaste-towear.medasset.org/en/socks/
16. Marine debris awareness poster https://web.whoi.edu/seagrant/outreach-education/marine-debris

12

Wood

12.1 Boats and Household

Wood is organic and can't be all that bad as marine debris, right? Wrong, opium is organic too, and that doesn't necessarily make it harmless. The fact that trees can stand for centuries if not millennia testifies to the inherent strength of wood. This material is very resistant to damage – just ask anyone who has ever hit a tree with a car. Wood also contains many natural "chemical warfare" substances that ward off insect pests or help keep neighboring, competing plants at bay. It is resistant in every sense of the word.

On the marine debris front, we can restrict ourselves to what is known as "processed wood," not to the weathered and worn driftwood representing the unadulterated remains of trees. In most fashioned products, wood's natural strength and durability are further enhanced by gluing it together into laminated layers, impregnating it with chemicals, or sealing it with coats of paint. The paints, varnishes, and lacquers used to preserve wood used in the marine environment (e.g., boat hulls) tend to be particularly hardy. In most cases they are intentionally toxic to prevent overgrowth by algae, mussels, tubeworms, sea squirts, and other so-called "fouling" organisms that weigh down and slow up vessels. Finally, many manufacturing processes give wood as a working material added strength through various structural design tricks supported by the liberal use of nails, screws, and heavy-duty staples.

All of these features are exemplified in boats, made of wood since before recorded history and only relatively recently being partly superseded by metal, fiberglass, plastic, and the like. You'll find wooden boats in every conceivable state of repair in and out of the water in every harbor as well as being operated by fishers and recreational captains along every stretch of coastline in the world. Why do boats become marine debris? They can accidentally break away from their moorings or docks. They may have been swept away by a tsunami. In emergencies they may have had to be abandoned at sea. People have also been known to "lose" their boats for insurance purposes. Finally, boats, like all possessions, have a certain life span. It is surprisingly expensive to maintain a vessel and even more so to refurbish one that has fallen into disrepair, not to mention paying for proper disposal/dismantling/salvaging. Moreover, boating often turns out to be less romantic than envisioned. One definition of sailing? Going nowhere very slowly, or arriving at great expense somewhere where you didn't want to land in the first place, cold, wet, hungry, and miserable… Owners may therefore simply abandon or run their old boats aground for someone else to find and deal with (potentially costing up to $50,000 to remove [1]). All this helps make boats and their components relatively common mega-litter. Derelict vessels pose multiple hazards. For one, they can damage coastal ecosystems when pushed and rubbed around about by waves (think coral reefs, seagrass beds, salt marshes). Hitting an abandoned boat at sea is a recipe for disaster. And they may contain fuel, oil, and other chemicals that pollute the water.

Processed wood as marine debris comes from many other sources as well. They include the fishing industry (crates to hold fish), industry (pallets to store and move goods), furniture, sports equipment, and children's toys. Even the food industry plays a role (ice cream sticks). Disposable wooden chopsticks alone consume 2 million cubic meters (70 million cubic feet) of timber each year. Globally, 1.4 billion people throw away 80 billion pairs every year, contributing to deforestation [2]. And, unsurprisingly, like any item produced in such numbers, they do land on beaches as well. What threats does wooden marine debris pose? First and foremost, wood floats. From an ecological perspective, wooden structures therefore "raft" organisms over large distances, introducing them as so-called alien species to faraway places. This includes the usual array of encrusting species (think barnacles) but also larger animals from lizards and snakes to rats and even larger mammals. As Charles Darwin recognized, this was once a natural phenomenon that helped to colonize remote or newly arisen islands and to trigger speciation. Today, however, the exotic arrivals on the disproportionately larger number of floating debris tend to wreak havoc among the native marine fauna and flora. When washing ashore, bulkier wood items can also scour the bottom and damage shallow marine habitats such as seagrass beds. Their back-and-forth movement in the surf zone, for example, damages coral reefs. For those who care somewhat less about things biological, there are clear hazards to humans as well. Heavy items such as beams and pallets floating on or just below the sea surface pose a considerable risk to smaller boats and their crews: you don't want to hit them at speed. Finally, on the beach, splinters and metal pieces – nails, brackets, and the like – are a threat to barefoot visitors. Most natural wood items should be left on beaches. They can play many beneficial roles, from stabilizing upper sandy beach areas to providing nutrients to dune vegetation. Some marine species – such as "shipworms," which are in fact bivalve molluscs – almost exclusively inhabit wood. Man-made – treated and processed – wood objects, however, should be collected during beach cleanups, with due consideration to their weight and with robust gloves to protect against splinters and nails. Among the weird wooden finds during recent international coastal cleanups: a piano and 15 brooms (handles) [3], but you'll find dozens of other items ranging from boats and their parts to fireworks sticks and pencils.

Fig. 12.1 Congratulations on bagging one of the "Big 10" marine debris items – a boat! And a colorful one at that! Caribbean, Grenada

Fig. 12.2 Many wooden boats are laid to rest rather than washed ashore – the equivalent of rusty old cars that some folks store in their yards with vague notions of future restoration. The rot on this hull and the empty propeller shaft point to "mega beach litter." Mediterranean, Greece

Fig. 12.3 Abandoned or derelict boats and "boat graveyards" are common in inlets and river mouths worldwide. Many are only a typhoon or flood away from becoming beach debris. Caribbean, Grenada

Fig. 12.4 This shattered bow indicates a violent end. Mediterranean, Turkey

Fig. 12.5 and Fig. 12.6 And here's the nearby stern to prove it. Mediterranean, Turkey. In older wrecks, the sturdiest parts – keel and ribs – are the nautical equivalent of animal skeletons sticking out of the desert sand. Mediterranean, Turkey

Fig. 12.7 The slight curvature, thick paint, and regular imprints of ribs are telltale signs of a hull plank. Mediterranean, Italy

Fig. 12.8 This crime scene wood is probably a former window in a small boat cabin. The clues: the recessed opening surrounded by closely spaced screw holes. Mediterranean, Turkey

Fig. 12.9 Wooden piece – probably to accommodate oars (note metal tube top center) – from a smaller wooden boat? Mediterranean, Slovenia

Fig. 12.10 And of course you'll find everything that can go overboard while boating, such as cushions that double as flotation devices. After all, who likes to sit on cold, wet wood or fiberglass. Pacific, USA

Fig. 12.11 A valuable boating-related marine debris item: a large fender used to protect moored yachts from harbor walls and neighboring vessels. Protective black stocking (an added luxury feature) extends the fender's life both on the boat and as marine debris. Mediterranean, Turkey

Fig. 12.12 OK beach sleuth, what does this have to do with boats? Good guess – it's the starter grip or handle from one of the millions of outboard motors plying the world's waterways. Pulled thousands of times with vehemence to start reluctant engines, the ropes tend to tear. Mediterranean, Turkey

Fig. 12.13 and Fig. 12.14 Large, heavy, weathered beams are long-lived marine debris items. Hitting one at speed with your boat can spell disaster. Caribbean, Cuba.
Wood, both large and small, tends to have nails or screws in it. According to Murphy's law, these always seem to be facing up. Even wearing tennis shoes may not protect you. Mediterranean, Turkey

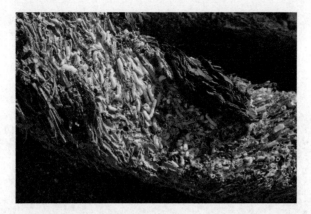

Fig. 12.15 Not immediately recognizable even to experienced beach detectives, this is actually a piece of wood completely perforated by burrowing molluscs (so-called shipworms). Such biological degradation requires lengthier submergence. Mediterranean, Turkey

Fig. 12.16 Paddle ball is a beloved beach game. The serial air-hole perforations were apparently the weak link in this product. Paint, lamination, and glue make many wood items less "organic" and prolong their marine debris life span.

Fig. 12.17 Why someone would leave behind a perfectly intact ping-pong racquet stumps even the best beach detective. Mediterranean, Turkey

Fig. 12.18 Brooms, rakes, and some children's toys actually still have wood parts. The stapled-on plastic grip of this shovel, even if flimsy, will survive on the beach much longer than the shaft. Atlantic, England

Fig. 12.19 Writing utensils are quite common marine debris items. Even a simple wood pencil is a classic example of a composite item: wood, paint, graphite, metal, rubber eraser. Every product can be made more eco-friendly: in this case a company has produced the first pencils made with 100% recycled cardboard and newspaper fiber. Atlantic, USA

Fig. 12.20 This wood ruler was made in China bearing imperial/US units and found its final resting place nicely embedded among cigarette butts on a Red Sea beach. Marine debris adds a new dimension to "the free movement of goods." Red Sea, Jordan

Fig. 12.21 If it comes with a fancy drink or with ice cream, you'll find it on tourist beaches. Mediterranean, Turkey

Figs. 12.22 and 12.23 Vacation, ice cream, and the beach. An inescapable combination. Wooden ice cream sticks are better than plastic and perhaps one of the more benign marine debris items. No matter how small the items, proud companies can't resist immortalizing themselves. Left, Pacific, Hawaii; right, Mediterranean, Greece

Fig. 12.24 Golf anyone? Beaches are, of course, the world's biggest sand trap. Atlantic, USA

Figs. 12.25 and 12.26 Paintbrushes, handles, and remnants often point to nearby harbors or dry docks. Keep an eye out for the paint cans as well – they're bound to be around somewhere! Mediterranean, Turkey

Fig. 12.27 Furniture is a staple on many beaches (see Chap. 8 (furniture)), but this wooden chair has passed the threshold to beach litter. Pacific, Panama

12.2 Pallets

Billions of tons of goods are moved every day, whether it be within shops, stores, and warehouses or across the sea in containers. Much of these goods are packed, handled, and delivered on pallets. Half a billion pallets are apparently being made each year: 2 billion are in use in the USA alone, another 3 billion in the European Union, untold billions worldwide. Is it any wonder that so many end up in the sea and on beaches? This makes them a standard item on international coastal cleanup data cards [4] and warrants a separate subchapter.

Pallets are practical because they can literally hold about a ton of goods (that's 1000 kilograms or a bit over 2200 pounds) and because they can easily be maneuvered by forklift vehicles and smaller hand-operated or electric pallet jacks. The products stacked up on them are typically secured with strapping bands and/or plastic sheeting (stretch or shrink wrap) – but that's another marine debris issue! Pallets can be made of plastic, metal, even paper – and certain parts (the blocks) of compressed wood chips or banana fiber – but most are made of wood. They come in many different "standard" sizes. This mirrors the fact that the world can agree on very little and that most efforts at standardizing anything at all fail for various historical, political, and/or vested commercial interests. The range of regional norms spans from the Grocery Manufacturers' Association pallet in North America to Europe's europallets. The International Organization for Standardization (ISO) sanctions six standard sizes, but various countries, military organizations, and industries use special sizes to suit their needs (e.g., shipping their motorcycles). Some are dimensioned to be able to pass through normal doorways, but most are considerably larger.

All wooden pallets cost something to make and ultimately involve cutting down trees. This makes them valuable! Some simpler pallets, however, are for one-time use and built of cheaper softwood, such as those often used to deliver originally packed furniture or a refrigerator to your door. These tend to be discarded along with the remaining packaging material after arrival. Others are made of hardwood and designed for repeated heavy-duty work. The costlier and more complex pallets (i.e., liftable from all four instead of just two sides) are typically reused or resold. These are also more likely to be stamped with company logos. Most pallets also have an International Plant Protection Commission (IPPC) certification stamped on or branded into two opposite sides (Fig. 12.28). Some companies color their pallets red (PECO, LPR), blue (CHEP), brown (IPP), or yellow (Yellow Pallet) for more instant recognition (and hopeful return); others variously color them depending on their customers.

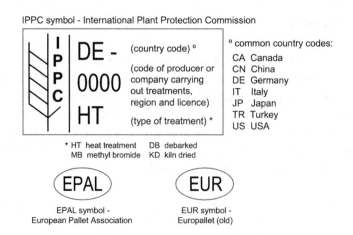

Fig. 12.28 Wood pallet markings. Upscaling heaven if you avoid those with the 'MB' symbol

If most pallets are reusable (or rentable) and valuable, why do they still end up littering our seas and beaches? Some are no doubt lost at sea, for example, when containers and other freight go overboard in storms. Damaged pallets may be discarded directly overboard or in harbors. Others are probably simply dumped with other municipal wastes into the sea.

Pallets must be very robust indeed to hold a ton of goods. Clever structural design along with heat or chemical treatment can delay breakage and slow down decomposition. Ultimately, however, even the sturdiest pallets deteriorate or break. A single broken plank or board can render a pallet useless. Minor damage, however, can and should be repaired. Most pallets can be dismantled, repaired, or reassembled by hand or by automated pallet repair machines. Improperly discarded pallets are where their marine debris career begins. Like all larger wood debris, pallets can damage shallow-water habitats and injure beach visitors (nails, splinters). Moreover, some can actually be toxic. The IPPC stipulates that most pallets shipped across national borders must be made of materials that cannot carry invasive species of insects and plant diseases. This typically means heat treatment: the wood must be heated to a minimum core temperature of 56 °C (132.8 °F) for at least 30 minutes [5]. Such pallets bear the initials HT near the IPPC logo. The second option is chemical treatment. These substances are by definition toxic – at least for many "lower" organisms. They include methyl bromide (check for the initials MD on the IPPC stamp), and their use has been partially banned (in the EU, for instance). A more harmless initial is KD (kiln dried), a gentle drying to remove moisture. Imported palletized goods may be additionally fumigated with pesticides (no special stamped initials), potentially leaving residues on the wood. Finally, harmful materials or chemicals from damaged products themselves may have spilt on and been absorbed by the pallet wood. Watch out for suspicious stains!

On beaches, all pallets and their remains should be collected. Watch out for projecting nails and splinters as well. Many pallets are partially buried in the sand. During beach cleanups you may need several people to pull them out and carry them off. Dig them free first and remove all sand to make them lighter. Watch your back and wear gloves!

What can you (or your company) do to reduce the number of derelict pallets? Reuse them as often as possible! One company claims that one tree is saved for every 10 pallets recycled (that's an acre of trees for every 200 pallets). What can be done with abandoned pallets and those removed from beaches? The possibilities appear to be virtually limitless [6]. Untreated pallets (i.e., those not bearing the ominous "MD" initials) can be used for firewood or for what appear to be 1001 unique, handsome, shabby-chic products. Used whole, they can be made into anything from tables to wine bottle holders. Stacked two to three high and topped with a fresh board or cushion makes them into robust indoor or outdoor seats. The individual boards can provide valuable material to build virtually any type of rustic furniture [7]. Just enter "reclaimed pallet" into your computer's search engine for endless suggestions and images of finished items. For this you will need to take the pallet apart. Enter "dismantle/disassemble pallet" into your computer for instructions and videos on the many clever ways to do this.

Fig. 12.29 Man-made "driftwood" like this pallet is marine debris and should be removed during beach cleanups! Caribbean, Panama

Fig. 12.30 Built to hold 1 ton of goods and to take punishment. In the surf zone, they can also mete out punishment to coral reefs, seagrasses, and other shallow habitats. Mediterranean, Turkey

Fig. 12.31 Watch your back and wear gloves when extracting pallets from the sand. Atlantic, Scotland

Fig. 12.32 Overturned pallet with fiberboard blocks. Exposed nails and splinters mean caution during beach clean-ups. The wood is a mecca for shabby-chic furniture and other upscaling projects. Atlantic, Scotland

Fig. 12.33 Pallets on rocky shores tend to be more battered and broken. Still solid enough to crush intertidal marine life, though. Caribbean, Guadeloupe

Fig. 12.34 Regularly spaced blocks tell beach detectives that this is a pallet board. These blocks can be tightly jammed between rocks by wave action and difficult to remove. Check stampings on the blocks for more information. Mediterranean, Croatia

Fig. 12.35 Battered, broken, and burned, but still recognizable as a pallet. Atlantic, Scotland

Fig. 12.36 Even the smallest pallet fragment at the crime scene should pose little identification problem for experienced beach detectives. Atlantic, Scotland

Fig. 12.37 Hard to believe that "the law of marine debris aggregation" also holds for pallets – you will rarely find only a single one! Check the internet for endless clever and profitable upcycling opportunities for pallets and their boards. Mediterranean, Croatia

Figs 12.38 and 12.39 Fishers need lots of handy crates to unload a good catch. These are often stacked high on exposed places onboard, helping explain why so many are lost. Mediterranean, Greece

Fig. 12.40 Rough seas can quickly turn loosely stacked fish crates into marine debris. Mediterranean, Greece

Fig. 12.41 Fish crates can get ugly fast even if nailed together and reinforced with brackets. A small mishap for a fisher, a giant leap for marine debris. Mediterranean, Greece

Fig. 12.42 Slat size and the rusty bracket are decisive clues pointing to a fishing crate of the type as in Fig. 12.41. Mediterranean, Greece

Figs 12.43 and 12.44 Flimsier wooden boxes used for fruits and vegetables are often stapled rather than nailed together. Exposed staples and splintered wood mean barefoot pain. Mediterranean, Italy, Turkey

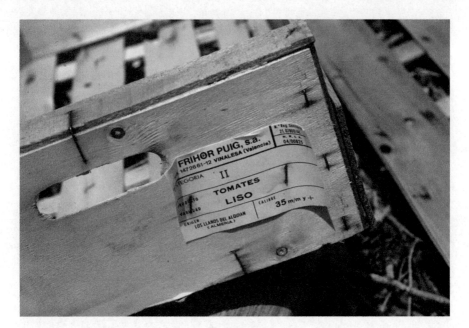

Fig. 12.45 This crate once held Spanish tomatoes. Note stapled construction, handgrip, and reinforced corners. Names, addresses, and telephone numbers on many marine debris items mean you can call up and complain if a certain item is over-represented on "your" beach. Mediterranean, Spain, Ibiza

References

1. The educator's guide to marine debris. Southeast and Gulf of Mexico. http://www.cosee-se.org/files/southeast/Marine_debris_guide1.pdf
2. Shaw A (2013) How wooden chopsticks are killing nature. http://www.ecopedia.com/environment/how-wooden-chopsticks-are-killing-nature/
3. Ocean Conservancy. https://oceanconservancy.org/wp-content/uploads/2017/06/International-Coastal-Cleanup_2017-Report.pdf
4. Ocean Conservancy. http://act.oceanconservancy.org/site/DocServer/ICC_Eng_DataCardFINAL.pdf?docID=4221
5. https://en.wikipedia.org/wiki/Pallet
6. http://www.1001palletideas.com/
7. https://www.1001motiveideas.com/home-decor/diy-ideas-recycling-wooden-pallets/

13

Paper

Paper sounds like it might be one of the more harmless types of marine debris. But any partying student who has ever thrown a roll of toilet paper can tell you that a little paper goes a long way to defiling the surroundings. A single newspaper peeled apart and scattered by the wind also leaves behind a considerably more chaotic impression than the neatly folded newsstand copy would have you believe. Importantly, most paper isn't as "organic" as most folks believe. Or did you think that paper money, for example, is really made entirely of paper? Although most paper is made of cellulose fibers derived from trees, numerous processes typically help it become more robust and long-lived. These include adding pigments or chemicals to avoid mold and light degradation or to make it particularly white. Laminations, wax, and admixtures of cotton or synthetic fibers can also help make paper more durable (and water resistant). Finally, various folding techniques and corrugation strategies add considerable strength. Getting to know such tricks is a favorite assignment for budding architecture students ("Can you design a paper/cardboard structure that can support 50 kg?"). Designers make robust furniture out of it, and some industrial pallets made of cardboard can support loads of up to 500 kg (enter "cardboard" and "pallet" into your computer's search engine).

All of this means that "paper" sticks around longer than most people might think. Everyone's emails these days seem to mention reducing paper use ("don't print this!") and saving trees. Our digital gurus never cease to wax poetic about how the electronic age is making paper redundant. Why then does so much paper pass through our hands every day, and why is there so much of it on the world's beaches?

Much of what you'll find on beaches is probably not washed ashore from faraway places but left by beachgoers. Again, that's you and me. Why? Think weekends and summer vacations – the beach is *the* place to relax and … read a hefty Sunday newspaper edition or all the magazine issues that have been piling up at home for weeks. The expressions "beach book" or "summer read" are no coincidence. And the hard fact (for the electronic industry) is that most laptops and other digital reading devices are simply not really designed for nasty beach conditions: too hot, too much sand, not to mention salt water – too risky. And the glare….

You won't be surprised about the many paper items folks bring to the beach: paper cups and plates, napkins, fast-food boxes, cigarette packages, and paper bags to put them all in, not to mention the packaging from all the freshly purchased beach paraphernalia ranging from sunglasses to flying discs and snorkeling equipment. Paper bags are a separate item on coastal cleanup data sheets, and paper makes up parts of many other items (e.g., tobacco packaging) [1]. On the bottom end (pun intended) of the scale is toilet paper – which any forward-thinking family will take to the beach for those inevitable emergencies. How much of just that one category of paper is produced? 83 million rolls (consuming 27,000 trees) *every day* [2]! And what type of "paper" is so valuable that it makes the "Big 10" list of debris you'll actually want to find? Correct, money: the bills that normally reside in your wallet but, on beaches, are typically more loosely stuffed into shorts or skimpy bathing suit pockets. You must have lost them on occasion, because I certainly find them every once in a while! This is one case where secondarily strengthened paper is a beachcomber's delight: the additives that make bank notes resilient also keep them in good shape, even in salt water.

Untold amounts of paper are produced and discarded worldwide (exceeding 400 million tons per year [3]) – despite all the digital hubris. Of all the municipal solid wastes produced in the USA, for example, paper makes up about one-third. "Municipal," simply put, again means mostly you and me. The newest source? The "cardboard crisis" involving miniscule items being sent in massive boxes, triggered by the exploding number of items we are purchasing online and that are delivered to our doorsteps in over-sized shipping boxes (additionally stuffed with paper at best, foamed plastic at worst). We are part of the problem and should be part of the solution. How about really not printing out that email? What about reading, and paying for, that newspaper online? And buying only recycled paper? On that note, paper was one of the first items to be recycled in enormous amounts and, today, makes up about half of all recyclables collected (by weight). You can also make an effort to ensure that almost all the paper used in your household (recycled printing paper) or taken to the beach is made partially of "postconsumer" waste and can be recycled once again. Recycling one ton of scrap paper can save anywhere between one and two tons of wood, ultimately protecting trees, especially praiseworthy if these are from old-growth forests. Importantly, recycling paper – rather than making paper from fresh pulp – can significantly reduce water and air pollution, cut energy costs, and ease the pressure on landfills.

On the marine debris front, paper is a tissue, I mean issue (see the following pages for examples). As problematic as it may be, however, it is clearly the lesser of two evils when compared with plastic. This means you should consider what plastic products you might be able to replace with (recycled) paper versions on the beach. This could include anything from straws and plates to hot drink sleeves (instead of foamed plastic finger protectors) to bags. Of course, paper bags are not the ultimate answer: at the 2012 international coastal cleanup, more than 298,000 paper bags were collected, putting this item on the "top 10" list of things found [4]. Think cloth bags! Some paper items can be upscaled too, with toilet paper rolls being an unexpectedly diverse example [5]. Of course, take any paper you bring to the shore back home, and collect all paper and cardboard items you encounter during beach cleanups.

Fig. 13.1 What should we all do on the beach? Drink plenty of fluids! Most "paper" cups are waxed or otherwise treated to hold fluids longer. Paper cups may be better than pure plastic, but stacking slows their decomposition. Mediterranean, Turkey

Figs. 13.2 and 13.3 Sticky slush ice cones or tubes are not the paper items folks like to take back home with them. Pacific, Hawaii; Mediterranean, Turkey

Fig. 13.4 This is an easy one for beach detectives. The tiny logo on top reads "pitch it," and someone clearly did. The caliper is opened to 1 centimeter, so this is a jumbo size. Pacific, Hawaii

Fig. 13.5 A new and exploding source of disposable paper cups: the coffee-to-go mania. Bring your own reusable mug! Mediterranean, Turkey

Fig. 13.6 Eating ice cream together on the beach is a wonderful treat. Plastic spoons, napkins, and a cone round off the beach debris mix these visitors left behind. Mediterranean, Turkey

Fig. 13.7 In the marine debris game, some companies probably wish they had sold their wares in plain brown wrapping. Pacific, California

Fig. 13.8 Although it may be the least of your injury-related concerns, most (beer) bottles have tenacious paper stickers. Mediterranean, Croatia

Fig. 13.9 Food wrappers (although not all are paper) are high on the "top 10" list of items collected during beach cleanups, directly after cigarette butts. On beaches, chocolate definitely melts in your hands before it melts in your mouth, leaving behind sticky wrappers. Atlantic, USA

Fig. 13.10 What's one of the most common pastimes at the shore? Reading. On windy beaches, most of the newspaper may never make it back home. It's enough to make you scream. Mediterranean, Turkey

Fig. 13.11 And all the great offers loosely tucked in your favorite magazine also tend to be blown away. Atlantic, USA

Fig. 13.12 Even cheap reads make for a lot of paper marine debris. Pacific, Japan

Fig. 13.13 Cardboard is a very common beach litter category. Here it was used to protect scantily clad butts from the gravel beach: a rather short-lived and short-sighted form of cardboard reuse. The plastic bag with left-behind garbage seals my judgment about these beach visitors. Mediterranean, Turkey

Figs 13.14 and 14.15 Freshly purchased beach items mean plenty of paper packaging and excessively bloated "instructions," often in a dozen languages including pictograms to service entire hemispheres. Unpack your purchase at the store and let them dispose of the paper. Mediterranean, Turkey

Fig. 13.16 Beach furniture packaging: made in China for Michigan and found on a Pacific beach in Hawaii. Don't you just love globalization?

Fig. 13.17 Most companies are so proud of their products; they're happy to divulge websites, email addresses, and telephone numbers. So feel free to give them a ring. This is a swimming fin insert. Mediterranean, Turkey

Fig. 13.18 Watch out for fishing-related paper: where there is packaging, the hooks might not be far off. Mediterranean, Turkey

Fig. 13.19 Fishing lure box, made in Finland, found on a beach in Mexico.

Fig. 13.20 Cigarette carton and newspaper reused to make a temporary paper fishing bait box. To paraphrase an old Star Trek saying: It's recycling Jim, but not recycling as we know it. Mediterranean, Turkey

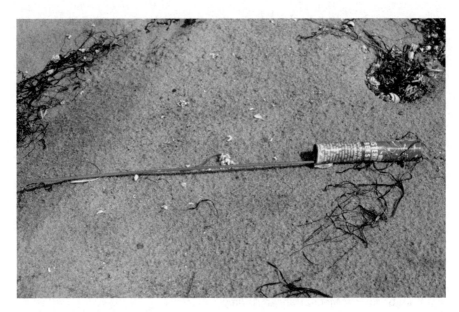

Fig. 13.21 Most folks think beaches are the perfect place to set off fireworks at night. They're wrong: think nesting sea turtles, sleeping birds, dry dune vegetation – and marine debris. Atlantic, USA

Fig. 13.22 Many warnings are printed on fireworks; unfortunately none of them pertain to marine debris. Paper meant to withstand pyrotechnics is robust and long-lived on beaches. Atlantic, USA

Fig. 13.23 As all pranksters know, a little bit of toilet paper can cover a lot of ground. Wherever you find it, "you know what" (see Chap. 14 (organic wastes)) is usually not far off. Mediterranean, Turkey

Fig. 13.24 "Paper" drink containers exemplify marine debris in featuring multiple components: plastic lamination, aluminum lining, adhesives, colors, not to mention sturdy plastic caps and seals. Atlantic, Morocco

Fig. 13.25 Snacking on the beach means lots of packaging. Good cardboard and metal bottoms and tops go a long way to making robust "paper" marine debris. Few folks heed all those cute little symbols (top left) about disposal and recycling. Mediterranean, Turkey

Fig. 13.26 Burning any marine debris, even paper, on the beach is a no-no fire hazard. Atlantic, USA

Fig. 13.27 "Paper" money, one of the "Big 10" marine debris items you like to find most on the beach! In this case, the large bill was a disappointment, worth less than US$ 1. Mediterranean, Turkey

References

1. http://act.oceanconservancy.org/site/DocServer/ICC_Eng_DataCardFINAL.pdf?docID=4221
2. https://en.wikipedia.org/wiki/Toilet_paper#Environmental_considerations
3. https://www.risiinfo.com/press-release/global-production-of-paper-and-board-hit-record-levels-in-2014/
4. Ocean Conservancy. Working for clean beaches and clean water: 2013 Report. https://oceanconservancy.org/wp-content/uploads/2017/04/2013-Ocean-Conservancy-ICC-Report.pdf
5. https://snapguide.com/supplies/roll-of-toilet-paper/

14

Organic Wastes

"Organic" normally sounds like a rather positive feature, for example, if you're into healthy food. But opium, snake venom, and plenty of other substances and materials are organic too, and that doesn't necessarily make them harmless or pleasant. Overall, I'm afraid you won't be too thrilled to encounter most of the marine debris items that are organic. Of course, many of the things beachcombers find on pristine seashores are organic, perfectly natural, and essential. Shorelines are exciting and high-biodiversity ecosystems with plenty of life (i.e., organic material). This includes everything from seagrasses and algae, clams, and mussels to snails and crabs. All their remains and remnants are also important for biological cycles and even beach sand formation. These items are the standard fare for traditional beach guides. If present in large amounts, however, they can also point to human impacts. Mass occurrences of jellyfish on beaches, for example, need not solely be a matter of unfavorable winds and currents: they can potentially reflect pollution of the water (eutrophication: over-enrichment with nutrients), ecosystem shifts due to overfishing, or the decimation of their main predators such as sea turtles! The same also holds true for excessive amounts of algae accumulated on beaches (algae need nutrients to grow and we are certainly providing them with more than enough). Unusual amounts of animal remains or washed-up fish also point to nothing good, for example, to recent mass mortalities due to insufficient oxygen in the water, also one of the symptoms of eutrophication.

In the framework of marine debris or beach litter, however, we are largely talking about "unnatural" organic material. Wastes that you and I are responsible for, either directly or indirectly, either intentionally or unintentionally. Organic debris can be understood to cover many different categories such as crude oil, wood, and paper, each of which is substantial enough as a pollution issue to warrant a separate chapter in this field guide. The present chapter concentrates on other solid items such as food remains, dung, and cadavers. Roughly one-third (1.3 *billion* tons) of the food produced annually for human consumption is lost or wasted worldwide. That value for root crops, fruits, and vegetables is 40–50% [1, 2]! Is it any wonder that some of this lands on our beaches?

One standard test of whether an item should be removed from the beach is whether you would want to step on it. I think most folks will agree that the answer to that question is "no" for almost all organic material (even if some of it is very soft indeed…). Moreover, many of these items decompose much more slowly than you might think: treated wood is meant to be tough, bones are hardy, and paper can be laminated or impregnated to increase durability. In the context of the present chapter, nature evolved the outer surface of every organism, whether animal or plant, to be resilient. Seeds and fruit skins, for example, are naturally resilient, and we often promote that with chemical treatments. This helps keep pests at bay, simplifies shipment, and increases shelf-lives in stores. It also means they stick around for a long time on beaches. Even a simple apple core may remain in the environment for 2 months, i.e., the whole vacation season in many parts of the world [3].

Importantly, any type of organic material will attract animals, typically scavengers ranging from birds and dogs but – depending on your geographic location – including raccoons, wild boar, jackals, and a range of more exotic beasts that you don't necessarily want to encounter on beaches either. Of course, getting too close to – or trying to take food away from – a hungry, feeding animal is never a good idea, as anyone with normally docile pets will confirm. Leaving food remains behind, even if in plastic bags next to full waste containers, is therefore not an option: animals *will* get to them and strew their contents everywhere. And animals are smart – they will return to such anthropogenic cornucopias regularly and teach other members of their population all the necessary tricks.

What do animals attracted to organic wastes leave behind on beaches other than ripped apart garbage bags and the strewn remains of gnawed food? Feces, dung, poop, droppings, crap, scat, manure, excrements – whatever coy name you want to give digestive products or however much you might wish to downplay them, animals that feed and "recycle" wastes on or near the beach don't necessarily yield a more palatable product. We may be used to dog poop on our city sidewalks and streets, but at least we have shoes on if we make a misstep. Beaches, however, are a barefoot environment. Importantly, it's much more than an esthetic issue. In this context it is useful to recollect what feces are actually composed of. I'll spare you the full details except to say that the answer isn't comforting. One component, beyond sloughed-off intestinal cells and food remains, is other organisms. These include a range of larger parasites and tinier bacteria – released by animals and humans alike. One gram of human feces may contain 10 million viruses, 1 million bacteria, 1000 parasite cysts, and 100 parasite eggs [4]. And all these organisms and life stages have evolved exceedingly clever strategies to become transmitted from one host to another. In fact, one of the key factors used to determine the quality of bathing waters and adjoining beaches is the amount of so-called fecal coliform bacteria. Higher values mean that the (necessary) bacteria from our digestive systems are not being properly removed by sewage facilities or that accidental wastewater discharges are occurring. This includes calculated losses, for example, overflows after heavy rains and flooding. The most visible manifestation in the organic waste category is feces themselves. On the one hand, this can be due to animals on the beach. It shouldn't be surprising that most tourist beaches therefore forbid taking pets to the beach, the world's largest litter box. On the other hand, it could mean your beach doesn't have adequate toilet facilities or visitors didn't make it there on time (see also "Toilets and Co." in Chap. 6). Another explanation is dysfunctional urban waste treatment or a sewage system consisting of a simple pipe opening a few hundred meters out at sea. Regardless of the source, you don't want your children unearthing brown heaps, sausages, or pellets when building their sand castle.

And while we are on the subject of animals, larger animals that have sought shelter on the beach should not be approached. In fact, no wildlife at all – whether on the beach or elsewhere – should ever be approached too closely. What is too close? Well, if the animal feels the need to stare at you directly, you are probably already too close. And please, please, please resist the urge to take a selfie with any wildlife: it definitely means you have gotten too close. Equally important, moribund or dead marine organisms that have washed up on the beach should never be touched. First of all, casual observers may be unable to distinguish sick – or even exhausted or sleeping – animals from dead ones. Stranded but living whales, for example, may take breaths in intervals of up to half an hour! Secondly, sea turtles, sea lions, seals, dolphins, and most other large marine species are protected by law and should never be annoyed, disturbed, inspected, manipulated, or moved, dead or alive, by the public. Never attempt to extricate entangled wildlife yourself. A panicked, thrashing animal on the beach can kill you. Many countries have "stranding networks": call them. Check out the many websites for more detailed instructions on how to behave and what positive action you can take [5]. Finally, sick, dead, or decaying organisms pose a serious health hazard, especially because some pathological agents such as bacteria can be transmitted from animals to humans. Inform your beach cleanup coordinator; call the local police or wildlife authorities. Removing dead whales, for example, can be a logistic nightmare. Some become bloated and blow up themselves [6]; others have actually been blown up using explosives, with unintended results [7]. If you are in a very remote area, you can gingerly (and with gloves) check dead birds for rings and sea turtles for tags. This information, sent along with a photo or two, can be very useful to authorities and researchers.

Fig. 14.1 Bananas and other peelable fruits are practical beach snacks but produce lots of resilient organic waste. And are often treated to ward off pests, prevent rot, and increase shelf life. The labels are nearly indestructible. Mediterranean, Italy

Fig. 14.2 and Fig. 14.3 There are many philosophies on how to cut a watermelon. All leave behind a mess. And all get ugly fast. Mediterranean, Turkey

Organic Wastes 317

Fig. 14.4 Corn is sold and eaten on many beaches worldwide. The remains are tough stuff. Mediterranean, Turkey

Fig. 14.5 Burning them doesn't solve the beach litter problem. After all, they are fire-resistant enough to make corncob smoking pipes! Mediterranean, Turkey

Fig. 14.6 and Fig. 14.7 Snacking on the beach is guaranteed to leave behind organic litter and packaging galore. Mediterranean, Turkey.
Dexterously nibbling on and spitting out sunflower (or other) seeds is a traditional pastime in many countries. Beach furniture is always a hotbed of organic waste production. Mediterranean, Turkey

Fig. 14.8 Food packaging, often still with organic contents, is at the top of the most common beach litter items reported by beach cleanups (after cigarette butts). Atlantic, Morocco

Organic Wastes 319

Fig. 14.9 On some beaches it's not difficult to scratch together a salad. Start off with a cucumber. Atlantic, Morocco

Fig. 14.10 and Fig. 14.11 And, of course, a sun-ripened tomato. Red Sea, Jordan ... then add the good parts of this pepper ... Mediterranean, Turkey

Fig. 14.12 Mix in something slightly chewy and a bit spicy. Mediterranean, Turkey

Fig. 14.13 Looks like we'll have to skip the tangy onion this time around. Red Sea, Jordan

Fig. 14.14 Mix it all together with a hand-picked, all-natural, beach-ripened lemon dressing. Mediterranean, Turkey

Fig. 14.15 Add a pinch of salt and bread: you're ready to go! Mediterranean, Turkey

Organic Wastes 321

Fig. 14.16 and Fig. 14.17 And for desert, some grapes. Mediterranean, Turkey. Somebody already beat you to this pear, though. Worldwide, 40–50% of all fruits and vegetables are discarded uneaten. Mediterranean, Greece

Fig. 14.18 Seeds like those of peaches or mangos, not to mention coconuts, are hardy and common beach litter. Mediterranean, Turkey

Fig. 14.19 Anything sweet and sticky dropped on the beach is bound to stay there. Not to forget the associated noise pollution – howling children. Mediterranean, Turkey

Fig. 14.20 Honestly, would you have salvaged this dropped lollipop and taken it back home for disposal? Mediterranean, Turkey

Fig. 14.21 Not everything organic is or was edible: the remnants of this "all-natural" broom are among the more benign forms of beach litter. Mediterranean, Turkey

Fig. 14.22 It may look appetizing, but it's beach litter nonetheless (straw). Nature made coconut husks to withstand years at sea. Caribbean, Guadeloupe

Fig. 14.23 Left behind or discarded by the dozen ("law of marine debris aggregation"). Beach detectives will recognize such "coconut drinks" even years later by the characteristic slice and hole. Caribbean, Guadeloupe

Fig. 14.24 Where you see animals on the beach, like these goats ... Mediterranean, Turkey

Fig. 14.25 ... or these cows being led to pasture, you will find their ... Mediterranean, Turkey

Fig. 14.26 ... footprints and ... Mediterranean, Turkey

Fig. 14.27 ... their droppings. Mediterranean, Turkey

Fig. 14.28 Cadavers (here a cow) and other bodily remains (mostly marine animals) are rather common on beaches. Mediterranean, Turkey

Fig. 14.29 Further south, the footprint on the right identifies these droppings on the beach as belonging to a camel. Red Sea, Jordan

Fig. 14.30 Webbed footprints? Beach detective response: water birds. Mediterranean, Italy

Fig. 14.31 Never touch living, dead, or dying animals, but you might gingerly (and with gloves) check and report whether a bird is banded (leg) or a sea turtle tagged (flipper). Take a photo or two if they are entangled in fishing gear or other plastic items. Mediterranean, Italy

Fig. 14.32 Beaches are the world's largest litter boxes. Dogs probably present the biggest problem: they leave behind lots of "dog doo," inspect and scatter garbage, not to mention dig up sea turtle nests. Mediterranean, Turkey

Fig. 14.33 The most dangerous footprint of all: that of humans. The official scientific term is "trampling," potentially damaging beach dunes and other sensitive coastal habitats worldwide. And wherever there are people, there is litter. Atlantic, USA

References

1. Rutten MM (2013) What economic theory tells us about the impacts of reducing food losses and/or waste: implications for research, policy and practice. Agric Food Sec 2:13 https://doi.org/10.1186/2048-7010-2-13
2. FAO (2011) Global food losses and food waste – extent, causes and prevention. FAO, Rome 30 pp
3. Marine debris is everyone's problem. https://web.whoi.edu/seagrant/wp-content/uploads/sites/24/2015/04/Marine-Debris-Poster_FINAL.pdf
4. International year of sanitation 2008: overview. https://www.unicef.org/…/E_-_IYS-UNICEF_Overview_10.doc
5. http://us.whales.org/issues/what-to-do-if-you-find-live-stranded-whale-or-dolphin
6. https://www.youtube.com/watch?v=8vVjxFdqoL8
7. https://www.youtube.com/watch?v=pAW4mntPM-w

15

Oil and Tar

Crude oil or tar in the water or washed ashore might not fit your notion of marine debris or beach litter in the strictest sense, but it is ugly, much of it is solid or at least firm, it poses a threat to wildlife and is hazardous to human health, and human activity put it there. "Black gold" is quite common on beaches and is definitely one of the most visible forms of marine pollution. All these features qualify it for inclusion in this field guide. Oil comes in many shapes and states: experts like to distinguish between oil sheens/oil films, liquid oil and oil slicks, "mousse" (a foamy water-and-oil emulsion), and tar balls [1]. The distinction is based mostly on age and decomposition stage but also reflects composition and even sea conditions. For beachgoers, however, the effect is pretty much the same: all these manifestations make your beach vacation miserable by keeping you out of the water and off the sand and scrubbing with cleansers if caught unaware. The effects of oil on wildlife are legendary. In extensive oil slicks, every organism that lands or feeds on the surface (birds) or has to come up to breathe (sea turtles, whales, and dolphins) will be covered. When birds attempt to clean themselves, they ingest the toxic oil. Sea turtles both large and small (hatchlings) are known to swallow tar balls (from an evolutionary perspective, everything floating in the sea was worth nibbling at), not to mention oiled beaches hampering their nesting [2, 3].

Billions of tons of oil are extracted every year, and about half of it is transported by tankers. Anywhere from 2 to 8 *million* tons are lost at sea every year [4]. Sound like a pretty imprecise range? The reason: no one really knows for sure, and it can change dramatically from year to year. When oil spills make the news, the media often report in a confusing array of units such liters, gallons, barrels, or tons – depending on who is doing the reporting and the degree of sensation/panic they want to cause or avoid. Environmental groups like to report in liters or gallons because that's a much larger number than the same volume in barrels or tons, preferably provided by industry. The sources of oil on the beach? Tanker accidents, tanker operations, oil platform operations, oil platform accidents and blowouts, coastal refineries, ship collisions, lost freighters and other commercial vessels (large freighters today carry more fuel oil than the loads of whole tankers in the early days), dry docking, scrapping of ships, bilge and motor oil from every boat type, industrial wastes, pipeline or storage tank leaks, urban runoff (look at the oil patches blotting almost every parking space when the cars are gone over the weekend), dumping, illegal oil changes, discarded oil canisters, and, in certain rare cases, natural seepage from the seafloor. Wow, that's quite a list! And who are the ultimate culprits? I hate to say it, but it's you and me as consumers. It's our demand that sends these incredible amounts of oil around the globe, and it's our appetite that fuels the spills and emissions. What kind of vehicle did you say you drive?

It's astounding that any beach can remain white for long. Infamous oil spills – the tankers Torrey Canyon, Amoco Cadiz, and Exxon Valdez and the oil platform accidents IXTOC 1 or Deepwater Horizon, to name but a tiniest selection – have guaranteed a steady string of major pulses of oil into the sea and onto our coasts over the last five decades. These are major media events and involve wrenching scenes of damaged land- and seascapes along with dead and dying seabirds and other marine life ranging from shore crabs to whales and dolphins. More important, much larger in overall quantity than major accidents, and more relevant in the context of this book are the untold, typically unreported steady spillages and losses associated with everyday work at sea and on land. This is typically the stuff that blackens our soles during beach vacations.

Crude oil comes from Mother Earth and is therefore natural and not so bad, right? Well, the deadly substances ejected by major volcanic eruptions and all the heavy metals (e.g., think mercury pollution) also stem from that same mother. That doesn't mean "natural" products are necessarily compatible with human biology. Crude oil consists mostly of hydrocarbons, i.e., hydrogen and carbon molecules in endless variations, but also contains sulfur, nitrogen, metals, and numerous other substances. Each oil field produces oil with a characteristic composition that can be chemically identified (useful in the event of a spill). Many of these components are toxic or lethal above certain amounts ("the dose makes the poison"). Finally, the refined oil products we buy from store shelves all have lots of additives to help them do their job better ("keeps your engine clean!"), none of which are principally designed with human or ecosystem health in mind.

This all means we have to fight oil pollution on every front. Unfortunately, no method of fighting oil spills will make your beach more beautiful or your vacation more enjoyable any time soon. Certainly not the various chemicals that are sprayed on oil to disperse it, or to "herd" it together, or stick to it and make it sink. Let's not even mention the biological methods ("let's spray it with special bacteria that can help break down the oil!"). The purely physical methods are also ecosystem killers. Cleaning rocky shorelines using high-pressure steamblasters or removing the top sand layers with heavy equipment simply compound the damage manyfold. Interestingly, the fine sand beaches we love so much are less sensitive than pebble beaches: that's because the oil tends to remain on the sand surface, where it can be more easily removed, whereas in coarse-grained beaches it seeps to greater depths where it cannot be removed and where it decomposes much more slowly [1]. The take-home message: major oil spills have enormous, decade-long environmental, legal, and financial repercussions.

If removing oil is so damaging, is the cleanup job better just left to Mother Nature? The answer in most cases is actually yes. All efforts to remove oil should focus on fighting it at sea. The primary goal: keep it from reaching the beach or other sensitive coastal ecosystems (think mangroves, coral reefs, marshes, and wetlands). In the environment, oil breaks down much like in a refinery: the small, volatile compounds evaporate first, and the larger ones successively separate out. On beaches, the process is temperature dependent. Sunlight and microorganisms help do the job, and both are much slower when it's cold or when the oil has seeped deeper into the beach. The final stage: the familiar asphalt-like tar balls. They can float on the sea surface for years; become overgrown with encrusting organisms, which makes them heavier; sink to the bottom (killing the organisms and making them lighter); rise to the surface again; and ultimately land on a beach somewhere – by the way, an up-and-down lifecycle also described for plastic items and even shoes.

What should you do if you encounter oil on your beach? That depends on the amount and type or stage of this "organic" material. Oil "bottles" are a separate item on international coastal cleanup data cards [5]. If it's a larger spill, the authorities will usually be in the know and the beach already the focus of an emergency plan. You don't have to have direct contact with oil to be at risk. The initial volatile compounds that evaporate from crude oil create toxic air pollution. That's why emergency cleanup crews wear protective suits and respiratory protection equipment. Very fresh oil is also flammable, which is why oil platform accidents often involve major explosions and fires. It also explains why leaking tankers have sometimes been bombed by the air force to burn off as much oil as possible before it can reach the shore [6]. After a chance encounter on the beach, you'll need chemical products or special wipes or towelettes to clean your oiled feet. If you are helping during a spill, take your oldest work apparel – you won't get around throwing away your oiled footwear and other clothing. During beach cleanups, wear disposable gloves if oil is present, and never open washed-up oil canisters and barrels – they are never empty to the last drop! Collecting and saving oiled birds and other wildlife are a job for professionals.

Fig. 15.1 Crude oil – fresh, soft, and shiny. Guaranteed to "tar" your feet or flip-flops. Pacific, USA

Fig. 15.2 Older oil becomes harder and duller. White barnacles mean time spent floating on the water surface. Pacific, USA

Fig. 15.3 If it's still flat, sticky, and accumulated at the waterline or high-tide mark, then it's usually part of an ongoing spill. Pacific, USA

Fig. 15.4 Stones and seagrass are embedded in this aged, dulled, hardened, and rounded tar ball. Caliper opened to 1 cm. Mediterranean, Greece

Fig. 15.5 Removing oil freshly washing ashore: a Sisyphus task requiring manual labor even if nicely visible against the white sand. Persian Gulf, Oman

Fig. 15.6 On some beaches, oil is apparently so common that tourists can be quickly warned with signs that are always on standby. "Wipes available at showers" in three languages – fighting the symptoms instead of the causes. Persian Gulf, Oman

Fig. 15.7 Fresh oil and tar accumulate at the high-tide mark and form a succession of ugly bands with the receding waters. Pacific, USA

Oil and Tar 335

Fig. 15.8 Oil spills attract the media and cleanup companies. Disposable protective clothing and gas masks are essential cleanup crew gear. Pacific, USA

Fig. 15.9 During beach cleanups, never open oil canisters. It's impossible to get the last drops of any oil out of any container! Pacific, Japan

Fig. 15.10 Oil barrels are often in poor shape, like this battered (but still closed and hazardous) drum on a boulder beach. Watch your back during coastal cleanups. Atlantic, Scotland

References

1. Clark RB (2001) Marine pollution, 5th edn. Oxford University Press, Oxford 248 pp
2. Gramentz D (1988) Involvement of loggerhead turtle with plastic, metal and hydrocarbon pollution in the central Mediterranean. Mar Pollut Bull 19(1):11–13
3. http://www.seeturtles.org/ocean-pollution/
4. Frid CLJ, Caswell BA (2017) Marine pollution. Oxford University Press, Oxford 268 pp
5. Ocean Conservancy. http://act.oceanconservancy.org/site/DocServer/ICC_Eng_DataCardFINAL.pdf?docID=4221
6. http://dragonsorb.net/torrey-canyon-worst-oil-spill-british-history/

16

Smoking

Smoking means different things to different people – a pleasure and hip activity, a sign of rebellion and maturity, a relaxant, an inconsiderate annoyance, a terrible scourge, and a medical epidemic – but for everyone, it is undisputedly a billion dollar industry, an ultimately deadly addiction, and a pollution source. It's hard to think of a more contentious lifestyle issue – besides dog poop on the sidewalk maybe. No matter how you look at it, however, discarded cigarettes and other smoking paraphernalia have no place on the beach, right? You'd think it would be counterintuitive to smoke while baking in the hot summer sun, but, hey, real smokers even light up on their breaks while fighting forest fires. Addicted is addicted.

Cigarette butts are the most frequent litter in the world and are by far the most common items you'll encounter on the beach. During beach cleanups, they consistently top the "dirty dozen" or "top 10" lists of marine debris by a huge margin every year. This warrants a separate heading "cigarette-related activities" on the cleanup data cards filled out by volunteers, which include cigarettes, cigarette filters, cigarette lighters, cigar tips, and tobacco packaging/wrappers [1]. The numbers collected during each annual 1-day beach cleanup hover around the 2 million mark [2]. This enormous amount is still trifling considering that an estimated 4.5 trillion of the 6 *trillion* – that's 6,000,000,000,000 – consumed worldwide every year end up improperly disposed of in the environment [3]. Whether left behind by beach visitors, thrown overboard from boats, or washed into the sea as runoff from streets via storm drains, rivers, and other waterways, many improperly discarded cigarettes ultimately end up on beaches. They are generally too small to be col-

© Springer International Publishing AG, part of Springer Nature 2019
M. Stachowitsch, *The Beachcomber's Guide to Marine Debris*,
https://doi.org/10.1007/978-3-319-90728-4_16

lected by beach-cleaning machinery and therefore tend to accumulate even on "cleaned" beaches.

Yes, but cigarette butts on the beach are rather harmless, right? Wrong again. Contrary to popular belief, cigarette butts are made less of paper than of thousands of cellulose acetate fibers, i.e., plastic. And, after smoking, these filters are further steeped in soaked-up tar, nicotine, heavy metals, and other carcinogenic residues. Cigarettes contain hundreds of additives, and over 4000 additional chemical compounds are created and released when a cigarette is smoked [4]. This makes them long-lived, miniature toxic waste bombs [5]. Just a few smoked butts (and even unsmoked!) per liter are sufficient to kill fish [4]. In studies that simulated rocky coastal tide pools, for example, the leached toxins from a few butts were enough to kill several species of snails [6]. Snails are normally a hardy bunch, so in practice, this means that the whole animal community in such pools can be wiped out. If you don't believe it, drop a cigarette butt or two into your aquarium at home and see how its denizens react.

Of course we are also talking about a major esthetic issue. Wading through stained and soggy cigarette remains along the waterline is unpleasant at best. Lying on a beach towel, you'll often smell the stale remains first, even if no butts are visibly exposed. These are often unexpectedly unearthed when absent-mindedly running you fingers through the sand or when your kids start digging trenches or building sand castles. Importantly, taking smoking behavior into consideration, you'll almost never find a single butt alone. A smoker spending the day on the beach is likely to leave behind a dozen or more souvenirs. This is the classical example of what I call "the law of marine debris aggregation": if you find one item, you probably don't have to look far to find a heap of their cousins. Some smokers think they are being considerate by pushing butts deep enough into the sand, so chances are you'll smell them before you see them. Other smokers leave their cigarette butts sticking upright out of the sand in densely packed groups – an ugly demonstration of imbecility. This means you'll find them accumulated in smoker-correlated hotspots (e.g., sunbeds). Many also tend to be accumulated at high-tide marks. Of course, all of them are only a storm away from becoming scattered all over the beach and back into the water.

Those who leave their butts behind should get their behinds butted. And smokers who leave their butts on or pushed into the beach also tend to leave behind their empty cigarette packs (and cartons), used matches, matchbooks and matchboxes, "empty" lighters, plastic filter mouthpieces, and other paraphernalia. All in all, this means a well-equipped but inconsiderate smoker is likely to deposit the entire spectrum of beach litter categories treated in this guide, including paper and cardboard, plastics, cellophane, adhesives, organics (tobacco), as well as aluminum foil and other metal components of lighters (not to mention the potentially explosive butane gas they contain at least in remnants). And let's not forget the cigar and pipe smokers who treat other beach visitors to their hobby. Folks who leave cigar butts on the beach probably have few qualms about leaving behind the wrappers or metal tubes in which the cigars came. And the newest craze? Electronic cigarettes with their liquid nicotine cartridges, atomizers, batteries, and LED lights. More than 13% of high school student smokers in the USA use e-cigarettes [7]. This is certain to add a new wave of problematic smoking-related items on beaches in the future.

Cigarette brands command intense loyalty, even if many of them are actually produced by the same conglomerates: although there are 100 or so notable brands, many are produced by what is sometimes referred to as the "Big 5" tobacco firms. Nonetheless, smokers like to think that they have made a personalized product choice and, in fact, the brand can sometimes tell you something about the motivations or lifestyle of a particular target group. A case in point: the "slim" cigarettes are designed to attract women. Aren't they just so graceful and glamourous! And, of course, there are cheaper and more expensive, filtered and unfiltered brands along with self-rolled cigarettes, all of which can reveal something about the individual smoker.

What features can beach sleuths work with? Lengths, for one, are a distinguishing factor: 70 mm ("regular size"), 84 mm ("king size"), 100 mm ("100's"), and even 120 mm ("120's") for those seeking the lengthiest enjoyment. Diameters also play a role: 9 mm is the norm, but sizes go down to 6–7 mm (slims) and even 5–6 mm (super/ultra-slims). Most importantly, cigarette companies are proud of their products, and you can be sure their name is on every single cigarette. The name is usually placed close to the filter and tends to remain visible even after the smoke, even if the smoker "goes all the way." Each brand additionally boasts its very own graphics. These are usually placed even further toward the lips (and thus even less likely to be burned away) in the form of one or more colored bands or rings, or at least a logo, between the tobacco rod and filter. The dedicated beach detective can identify almost any cigarette butt based on these features. Of course, each smoker leaves behind enough DNA to make personal identification based on a cigarette butt possible these days, but we don't need to go there.

How long do cigarette butts remain on the beach? Anywhere from 1.5 to 10 years [8]. This is because the filters themselves are non-biodegradable plastic and are additionally soaked with toxic smoking residues. Small plastic filter inserts or other patents that contain activated charcoal elements or "cooling" menthol capsules to treat the smoke won't disappear any time soon either.

Beyond being an esthetic issue, smoking is hazardous and downright lethal – to humans, animals, and even plants (you have no doubt witnessed what happens to potted plants when these are used as ashtrays). The toxic residues in discarded filters can poison and kill marine organisms, such as those living in the tide pools mentioned above. Then there is the threat of forest fires by cigarette butts flicked into the dry dune and back-beach vegetation during the hot summer months. If the damage to nature doesn't faze you, then smoking is deadly to human beings too. It has been associated with diabetes, liver and colorectal cancer, lung cancer, and heart disease. The estimate of premature deaths due to smoking is 5 million per year worldwide, making smoking an important cause of global mortality [9]. By 2050, that number is projected to be 3 million in China alone. If you're not worried about yourself, how about others? Like the number of miscarriages caused by smoking, or the number of newborn babies admitted to intensive care units because of low birth weight related to smoking parents? More generally, second-hand smoke is second-hand smoke, even if it's on the beach. One recent Italian study revealed that passive smoking on the beach exposes visitors to air that is twice as polluted as at a local traffic circle [10].

How can this marine debris problem be solved? As the saying goes "Only you can prevent forest fires" – and only you can help reduce smoking-related beach litter! Premature and horrible death should be enough to prompt strategy number one: please quit smoking. On site, how about more and better (non-flammable) refuse containers on beaches. And stiffer fines for littering. Some ocean resort towns have begun banning smoking on the beach [10]. On the industry side, the tobacco firms could upscale their research into biodegradable filters or consider a deposit system for returned filters. Some thought is being given to recycling by adding other plastics to the cellulose acetate filters to enable producing industrial (not household – too much nicotine) plastic products, but don't count on this reducing cigarette marine debris any time soon. Banning filtered cigarettes is also an option, especially because filters apparently do not reduce the inhaled contaminants much (versus unfiltered products), lulling many smokers into believing that the filters help protect their health (and discouraging them from attempting to quit). Finally, you can't leave the protection of the chicken coop to the foxes. On the smoking front, this calls for going a level above the industry. In the USA, one state (Hawaii) has taken the lead by raising the legal age for smoking to 21 years. The pallet of other measures ranges from making smoking prohibitively expensive to increasing smoking restrictions, banning advertising, eliminating special aromas and flavors, providing even more dire warnings and shock photos on cigarette packs, and tasking tobacco firms with the costs for studies on

Fig. 16.1 Cigarette butts: by far the most frequent litter item on beaches and always number 1 in every "top 10" beach cleanup list. Anyone who leaves their butts behind deserves to have their behind butted. Pacific, USA

Fig. 16.2 Being proud of your product means having your name clearly visible on every single one (on the part that doesn't burn off, of course). Pacific, Japan

Fig. 16.3 Here's one "with the works": name, graphics, unique bands, and rings. Pacific, USA

Figs 16.4 and 16.5 For beach detectives the lipstick marks are a dead giveaway – the culprits at this crime scene were women. Mediterranean, Turkey; Pacific, Japan

Fig. 16.6 Some folks apparently can smoke and chew gum at the same time without getting confused.

Fig. 16.7 At some point, the paper covering is washed off, exposing the plastic fibers that make up most of the cigarette butt. Atlantic, USA

Fig. 16.8 Additional plastic in the butt to make for "cooler" smoking. Prolonging the marine debris lifespan of cigarettes (already up to 10 years) and creating...instant microplastic. Mediterranean, Turkey

Figs 16.9 and 16.10 Plastic mouthpieces are toxic, built to take the heat and the bite: perfect ingredients for long-lived marine debris. Mediterranean, Turkey; Pacific

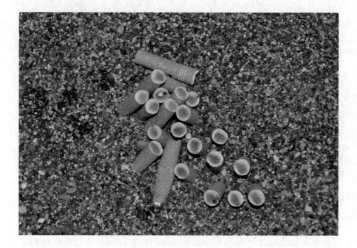

Fig. 16.11 The "law of marine debris aggregation" exemplified: you'll almost never find an isolated butt. Hours of smoking pleasure as a marine debris factory! Mediterranean, Turkey

Fig. 16.12 and Fig. 16.13 And where did smokers learn this dastardly "stick-it-in-the-sand" behavior? Maybe from sand-filled ashtrays such as found in posh tennis and golf clubs. USA.
Or maybe at sophisticated highway rest stops like this one. Croatia

Fig. 16.14 Throwing butts into the water may help prevent forest fires but will snuff out all this tide pool's inhabitants. Drop one into your aquarium at home and see what happens. Mediterranean, Croatia

Smoking

Fig. 16.15 Cigarette butts tend to wash up and accumulate along the high-tide line. Mediterranean, Turkey

Fig. 16.16 and Fig. 16.17 You really get to appreciate their actual numbers on black sand beaches (and check out the straws too!). Atlantic, Teneriffe.
Cigarette 'soft packs' consist of at least four components: outer plastic foil, paper-pack, aluminum foil insert, and adhesives.

Fig. 16.18 A pack a day means a pack a day thrown away. Paper, foil, color pigments, all enclosed in cellophane: a classical example of multicomponent marine debris. Atlantic, USA

Figs 16.19, 16.20 and 16.21 Dire warnings, regardless how boldly black-framed, how graphic, apparently have little effect. Photos of a healthy versus smoker's lungs (center) may be horrifying, but pointing out reduced virility/fertility (right) may make even the most hardcore smoker think twice! Mediterranean, Turkey

Fig. 16.22 Loggerhead sea turtle hatchling making its way around cigarette paraphernalia. They can be permanently trapped in small items such as cups and plastic bags. Marine debris is a conservation issue both on land and at sea! Mediterranean, Turkey

Fig. 16.23 and Fig. 16.24 Beach detective interpretation of these unsmoken cigarettes? Two smokers were surprised by a large wave, leaving behind only water-stained beach litter as testimony to their existence? Mediterranean, Turkey.
Cigars are best locked away in humidors. Mediterranean, Turkey

Fig. 16.25 Folks who discard their cigar butts on beaches probably don't think twice about ditching the metal packaging they come in. Caribbean, Guadeloupe

Fig. 16.26 And no metal packaging ages nicely as marine debris. Pacific, USA

Fig. 16.27 Sometimes you'll find the complete set: a delightful find … for a smoker. Mediterranean, Turkey

Fig. 16.28 Cigarette lighters – a mixture of plastic, metal, and flammable liquids. Miniature bombs one might say. The labels have many admonitions regarding exposure to sunlight and children, but nothing about leaving this "disposable" item on the beach.

Fig. 16.29 and Fig. 16.30 Like matchbooks, cigarette lighters are often freely distributed as advertising. Beach detectives have no trouble finding out where to complain. Mediterranean, Turkey.
Product safety considerations mean solid build and long-lived marine debris. Broken lighter with rusting metal cap (shroud), internal mechanism, and plastic "dip tube." Mediterranean, Turkey

Fig. 16.31 This crushed lighter contains small dice – certainly not lucky for the next one who steps on it barefoot. Mediterranean, Turkey

Fig. 16.32 See who can collect the most in 15 minutes along your favorite shoreline. Or "pick" other common litter items such as cigarette butts, straws, clothespins, cotton swabs, and plastic bottle caps…, and make a family competition out of it. Never another boring day at the beach! Mediterranean, Turkey

Fig. 16.33 and Fig. 16.34 Some traditionalists still use wooden matches. Made in Sweden, discarded in Greece. Free trade at its best. Mediterranean.
Have you ever seen anyone actually *not* discard a once-lit match? Atlantic, USA

Fig. 16.35 And, finally, let's not forget the ashtrays to complete the smoking paraphernalia. Look under Chap. 2 (glass) to find an equally attractive beach litter model. Mediterranean, Turkey

hazards and compliance [11]. Call your local political representatives and give them a piece of mind.

References

1. Nature Conservancy. http://act.oceanconservancy.org/site/DocServer/ICC_Eng_DataCardFINAL.pdf?docID=4221
2. Nature Conservancy. https://oceanconservancy.org/wp-content/uploads/2017/04/2016-Ocean-Conservancy-ICC-Report.pdf
3. http://www.cigwaste.org/
4. Slaughter E, Gersberg RM et al (2011) Toxicity of cigarette butts, and their chemical components, to marine and freshwater fish. Tob Control 20(Suppl 1):i25–i29 http://tobaccocontrol.bmj.com/content/20/Suppl_1/i25
5. Bonanomi G, Incert G et al (2015) Cigarette butt decomposition and associated chemical changes assessed by 13C CPMAS NMR. PLoS One 10:e0117393. https://doi.org/10.1371/journal.pone.0117393

6. Booth DJ, Gribben P, Parkinson K (2015) Impact of cigarette butt leachate on tidepool snails. Mar Pollut Bull 95(1):362–364 https://doi.org/10.1016/j.marpolbul.2015.04.0044
7. Smoking trends (2015) Seven days: the news in brief. Nature 520:23
8. Marine debris is everyone's problem. https://web.whoi.edu/seagrant/wp-content/uploads/sites/24/2015/04/Marine-Debris-Poster_FINAL.pdf
9. Ezzati M, Lopez AD (2003) Estimates of global mortality attributable to smoking in 2000. Lancet 9387:847–852 https://doi.org/10.1016/S0140-6736(03)14338-3
10. http://www.bibione.com/en/events-category/2327-bibione-smoke-free-beach-en
11. Novotny TE, Lum K et al (2009) Cigarette butts and the case for an environmental policy on hazardous cigarette waste. Int J Environ Res Public Health 6:1691–1705. https://doi.org/10.3390/ijerph6051691

Index

A

Air mattress, 201, 258
Aluminum cans, 51, 52
Antarctic, 67, 75
Anthropocene, 94
Apparel, 1, 2, 11, 31, 219–246, 332
Apple, 314
Aquaculture, 276
Ashtray, 31, 96, 100, 340, 344, 351

B

Bag, 9, 13, 14, 23, 24, 26, 33, 90, 94, 144–146, 149, 307
Balloon, 1, 12, 19, 89, 128–134, 181
Banana, 10, 291, 316
Band-aids, 200
Barnacle, 5, 9, 53, 83, 106, 117, 227–229, 238, 250, 265, 271, 274, 280, 332
Barrel, 22, 27, 51, 61, 155, 156, 177, 192, 329, 332, 336
Baseball cap, 244, 245
Bathroom sink, 188
Battery, 69, 72, 212
Beach ball, 256, 257
Beach furniture, 69, 201–206, 211, 308, 318
Beanbag, 166
Bicycle, 74
Bicycle lock, 75
Bicycle tires, 85
Bikini, 223
Biohazard, 21, 27, 241
Bird, 14, 44, 50, 54, 78, 90, 147, 179, 265, 269, 310, 314, 315, 326, 329, 332
Boat, xii, 8, 9, 16–20, 34, 44, 50, 56, 66, 71, 77, 80, 81, 91, 183, 184, 202, 221, 264, 279–282, 286, 330, 337
Box spring, 64, 210
Bra, 223
Bucket, 13, 17, 21, 26, 118, 121
Bulldozer, 75
Buoy
    foamed plastic, 159–162
    plastic, 88, 94, 108

C

Cadaver, 90, 314
Camel, 325
Canister, 1, 59, 60, 106, 114–121, 330, 332, 335
Cap, 33, 44, 54, 72, 97, 98, 103, 104, 110, 194, 196, 245, 246, 349
Cardboard, 10, 155, 162, 163, 177, 288, 301–303, 307, 311, 338
Carpeting, 88
CDs, 141
Chair, 98, 201, 202, 290
Chopsticks, 24, 280
Cigar, 337, 339, 347
Cigarette, 94, 337–340, 342, 343, 346–350
Cigarette butt, 15, 22, 32, 55, 96, 97, 120, 289, 306, 318, 337–341, 345, 350
Cigarette carton, 309
Cigarette packages, 302
Clamshell, 160, 162, 163
Clothes pins, 158, 176
Clothing, 1, 88, 93, 219–226, 244, 266, 332, 335
Coconut, 321, 323
Coffee-to-go, 24, 159, 304
Comb, 174
Commercial fisheries, 16–18, 44, 168, 264, 272
Compact music cassette, 141
Computer, 160, 212, 213, 215
Condom, 133, 181, 182
Contact lens, 181
Corn, 93, 317
Corrosive, 20, 27, 50
Cotton swab, 171, 175, 176, 350
Cow, 324, 325
Cruise line, 19, 156
Cucumber, 319
Cup
  paper, 24, 302–304
  plastic, 93, 100, 118, 134, 346
Cutter knives, 138

D

Danube, 93
Deodorant, 173, 180
Diaper, 172, 173, 179
Diving fins, 254
Diving masks, 254
Dog, 14, 15, 65, 147, 179, 231, 314, 326, 337
Doll, 121, 127
Dung, 314

E

Engine, 9, 18, 87, 91, 145, 330
Environmental hazard, 27
EPAL, 292
EU, 94, 144, 212, 293
Expanded polystyrene foam (EPS), 89, 159, 162
Explosive, 27, 56, 263, 315, 338

F

Fake grass, 144
Feces, 314, 315
Fender (boat), 80, 81, 285
Fender (car), 72
Fin, 90, 250, 251, 254, 308
Fireworks, 12, 128, 281, 310
Fish hooks, 26, 64, 265, 266
Fishing crate, 264
Fishing lure, 90, 268, 309
Fishing net, 18, 19, 89, 91, 160, 161, 220, 265, 266, 273
Flammable, 27, 331, 348
Flip-flop, 2, 26, 228, 234–241, 332
Flipper, 90, 148, 226, 251, 269, 326
Foam peanuts, 160, 167
Food, 10, 13, 16, 23, 24, 33, 45, 58, 88, 93, 134, 136, 145, 151, 159–161, 163, 220, 280, 306, 313, 314, 318
Footprint, 5, 220, 324, 325, 327
Fruit, 10, 314, 316, 321

Furniture, 1, 11, 25, 69, 134, 201–213, 280, 290, 291, 301, 308, 318

G

Gas cartridge, 50, 65, 84
Glass bottle, 24, 33–43, 97, 106
Glass floats, 43
Glove, 1, 2, 17, 26, 50, 52, 60, 172, 181, 183, 192, 193, 198, 221, 241–244, 265, 315, 326, 332
Glow sticks, 270
Goats, 324
Golf, 290, 344
Grapes, 321
Great Pacific Garbage Patch, 7
Grill, 49, 63, 212
Guinness World Record, 32

H

Hair bands, 173
Hairbrush, 173
Hair clips, 173
Hard hat, 21, 244, 246, 247
Harley-Davidson, 50
Hat, 1, 2, 13, 21, 244–248
HDPE, 117
Heel, 231, 236, 251
Helmet, 159, 161
HT, 292
Hubcaps, 68, 72

I

Ice cream, 13, 143, 152, 220, 289, 305
Ice packs, 142
Infusion apparatus, 199
International Coastal Cleanup, 22, 26, 62, 63, 74, 77, 134, 175, 192, 202, 212, 213, 217, 281, 291, 303, 331

International Plant Protection Commission (IPPC), 291, 292
International Whaling Commission (IWC), 90, 265
Irons, 212

K

Ketchup bottle, 153
Kitchen sink, 1, 67, 188

L

Label, 1, 25, 33, 34, 58, 61, 97, 104, 110, 114, 116, 241, 316, 348
Land mine, 50, 65
LDPE, 89
Lemon, 320
License plate, 68, 73
Life buoy, 160, 169
Life-saver, 160
Light bulb, 1, 17, 21, 31, 44–48, 105
Lighter, 23, 97, 158, 176, 216, 293, 331, 337, 338, 348, 349
Lollipop, 322

M

Mariculture, 276
Marine Pollution Bulletin, 7
Marine pollution (MARPOL), 21
Mask, 13, 148, 252–254, 335
Matches, 338, 350
Mattress, 64, 201, 210, 258
MB, 292
Medication, 197, 198
Microbeads, 172
Microplastic, 137, 139, 140, 158, 163, 166, 168, 255, 343
Military, 16, 19–21, 78, 291
Money, 9, 13, 18, 33, 51, 68, 91, 301, 302, 312

Motorcycle, 50
Music cassette, 141, 142

N
Nail, 50, 171, 177, 230, 231, 279, 281, 286, 292, 293
Needle, 2, 192, 193, 196
Neon light, 47
Newspaper, 34, 162, 288, 301, 302, 306, 309
Nozzle, 135, 136
Nurdles, 91, 92

O
Ocean Conservancy, 22, 26, 27, 44, 99
Oil, vii, 1, 8, 11, 16, 20–21, 44, 59, 61, 68, 78, 87, 94, 280, 314, 329–336
Oil changes, 119, 330
Oil drum, 51, 336
Oil spill, 329–331, 335
Onion, 320
Oxidizing, 27, 53
Oxygen mask, 199

P
Packing noodles, 160, 162, 167
Packing peanuts, 160, 167
Paddles, 262
Paint brushes, 139
Pallet, 22, 25, 280, 281, 291–293, 301, 340
Peaches, 321
Pear, 321
Pellet, 50, 91–93, 155, 156, 162, 166, 269, 315
Pen, 134
Pencil, 281, 288
Pepper, 319

PET, 88
Pills, 191, 197
Ping-pong racquet, 287
Plank, 284, 292
Plastic bottle, 6, 94, 97, 98, 105, 108, 136, 350
Plastic cup, 93, 100, 118
Plastic fork, 125, 134
Plastic knife, 91, 145
Plastic plates, 134
Plastic spoons, 305
Plastiglomerate, 94
Plastisphere, 93
Pliers, 63
Plumbing, 185
Plunger
   syringe, 193, 195
   toilet, 177
Polyethelene (PE), 88, 172, 259
Polypropylene, 117
Polystyrene, 159–163
Polyvinyl chloride, 88
Pool noodles, 259
Popcorn, 162, 167
PP, 88
PS, 88, 159
Pull tab, 54
PVC, 88, 241

R
Radioactive, 27, 191, 192
Recycle, 24
Recycling codes, 88, 89
Refrigerator, 49, 62, 202, 212, 213, 291
Refuse, 23
Remote control, 214
Repair, 24
Rethink, 23
Reuse, 24
Rhine, 93
Rope, 119

Rubber band, 182
Rubber boot, 233
Rug, 201, 211
Ruler, 289

S

Sandal, 76, 226, 234, 238, 240
Sanitary napkins, 178
Scrubber, 91, 172
Sealant tubes, 136
Selfie, 217, 315
Shaver, 94, 173
Shaving brush, 171, 174
Shipworms, 281, 287
Shipwrecks, 17, 66
Shoe, 2, 76, 121, 192, 193, 225–234, 240, 286, 314, 331
Shoe sole, 220, 230
Shopping carts, 62
Shotgun shell, 1, 2, 50, 155, 156, 176, 269
Shower head, 188, 248
Sidewall (tires), 79
Sieves, 121, 123
Six-pack holder, 89, 96, 99, 113
Snorkel, 13, 91, 148, 249–252, 254, 265, 269, 302
Soap tray, 188
Sock, 220, 225, 266, 276
Space, 8, 16, 19, 77, 113, 330
Spray can, 49, 56, 57, 173
Squirt gun, 126
Strapping band, 89, 90, 150, 291
Straw, 24, 96–99, 102, 103, 303, 323, 345
Styrofoam, 94, 159–162, 196, 261
Sunbed, 201–203, 205, 210, 338
Sunflower seed, 318
Surfboard, 250, 260, 261
Swim noodles, 259
Swimming goggle, 255, 256
Syringe, 2, 26, 89, 94, 191–196

T

Table, 49, 201, 202, 293
Take-away, 314
Tampon applicators, 177
Tampon, 177, 178
Tape measure, 141
Tar, 1, 11, 20, 103, 329–332, 334, 338
Tar ball, 20, 329, 331, 333
Telephone, 61, 161, 213, 308
Tetrapack, 109
Tire, 22, 27, 69, 76, 77, 80, 82–85
Toilet, 10, 16, 177, 183–189, 202, 302, 303, 311, 315
Toilet bowl cleaner, 183
Toilet bowl sanitizer, 183, 187
Toilet brushes, 183, 185, 186
Toilet paper, 183, 187, 301–303, 311
Toilet paper rolls, 187, 303
Toilet seat, 183, 184
Tomato, 319
Toothbrush, 171–173, 175
Toothpaste, 152, 172, 180
Toy, 1, 94, 121–129, 134, 280, 288
Toy truck, 121
Trampling, 5, 327
Truck, 68, 69, 76, 79, 121, 124, 125
Tubeworm, 83, 139, 156, 164, 233, 238, 243, 279

U

Umbrella, 13, 148, 201, 203, 207, 208, 219
Underwear, 224, 225
United Nations Environment Programme (UNEP), 6, 10, 20
Upcycle, 25, 295

V

Vegetables, 314, 321
Vehicle, 5, 34, 49, 69–73, 75, 83, 168, 291, 330

W
Wastewater treatment plants, 172
Watch, 138
Watermelon, 316
Whaling station, 67, 90, 265
Window pane, 31
Woggles, 259
Wooden pallet, 25, 291

Y
Young, N., 193

Z
Zappa, F., xv, 219